Fundamentals of Electronics

Book 4

Oscillators and
Advanced Electronics Topics

Synthesis Lectures on Digital Circuits and Systems

Editor
Mitchell A. Thornton, *Southern Methodist University*

The *Synthesis Lectures on Digital Circuits and Systems* series is comprised of 50- to 100-page books targeted for audience members with a wide-ranging background. The Lectures include topics that are of interest to students, professionals, and researchers in the area of design and analysis of digital circuits and systems. Each Lecture is self-contained and focuses on the background information required to understand the subject matter and practical case studies that illustrate applications. The format of a Lecture is structured such that each will be devoted to a specific topic in digital circuits and systems rather than a larger overview of several topics such as that found in a comprehensive handbook. The Lectures cover both well-established areas as well as newly developed or emerging material in digital circuits and systems design and analysis.

Fundamentals of Electronics: Book 4 Oscillators and Advanced Electronics Topics
Thomas F. Schubert, Jr. and Ernest M. Kim
2016

Fundamentals of Electronics: Book 3 Active Filters and Amplifier Frequency
Thomas F. Schubert and Ernest M. Kim
2016

Bad to the Bone: Crafting Electronic Systems with BeagleBone and BeagleBone Black, Second Edition
Steven F. Barrett and Jason Kridner
2015

Fundamentals of Electronics: Book 2 Amplifiers: Analysis and Design
Thomas F. Schubert and Ernest M. Kim
2015

Fundamentals of Electronics: Book 1 Electronic Devices and Circuit Applications
Thomas F. Schubert and Ernest M. Kim
2015

Applications of Zero-Suppressed Decision Diagrams
Tsutomu Sasao and Jon T. Butler
2014

Pragmatic Electrical Engineering: Systems and Instruments
William Eccles
2011

Pragmatic Electrical Engineering: Fundamentals
William Eccles
2011

Introduction to Embedded Systems: Using ANSI C and the Arduino Development Environment
David J. Russell
2010

Arduino Microcontroller: Processing for Everyone! Part II
Steven F. Barrett
2010

Arduino Microcontroller Processing for Everyone! Part I
Steven F. Barrett
2010

Digital System Verification: A Combined Formal Methods and Simulation Framework
Lun Li and Mitchell A. Thornton
2010

Progress in Applications of Boolean Functions
Tsutomu Sasao and Jon T. Butler
2009

Embedded Systems Design with the Atmel AVR Microcontroller: Part II
Steven F. Barrett
2009

Embedded Systems Design with the Atmel AVR Microcontroller: Part I
Steven F. Barrett
2009

Embedded Systems Interfacing for Engineers using the Freescale HCS08 Microcontroller II: Digital and Analog Hardware Interfacing
Douglas H. Summerville
2009

Designing Asynchronous Circuits using NULL Convention Logic (NCL)
Scott C. Smith and JiaDi
2009

Fundamentals of Electronics: Book 4 Oscillators and Advanced Electronics Topics
Thomas F. Schubert, Jr. and Ernest M. Kim

ISBN: 978-3-031-79885-6 paperback
ISBN: 978-3-031-79886-3 ebook

DOI 10.1007/978-3-031-79886-3

A Publication in the Springer series
SYNTHESIS LECTURES ON DIGITAL CIRCUITS AND SYSTEMS

Lecture #50
Series Editor: Mitchell A. Thornton, *Southern Methodist University*
Series ISSN
Print 1932-3166 Electronic 1932-3174

Fundamentals of Electronics
Book 4
Oscillators and Advanced Electronics Topics

Thomas F. Schubert, Jr. and Ernest M. Kim
University of San Diego

SYNTHESIS LECTURES ON DIGITAL CIRCUITS AND SYSTEMS #50

ABSTRACT

This book, *Oscillators and Advanced Electronics Topics*, is the final book of a larger, four-book set, *Fundamentals of Electronics*. It consists of five chapters that further develop practical electronic applications based on the fundamental principles developed in the first three books.

This book begins by extending the principles of electronic feedback circuits to linear oscillator circuits. The second chapter explores non-linear oscillation, waveform generation, and wave-shaping. The third chapter focuses on providing clean, reliable power for electronic applications where voltage regulation and transient suppression are the focus. Fundamentals of communication circuitry form the basis for the fourth chapter with voltage-controlled oscillators, mixers, and phase-lock loops being the primary focus. The final chapter expands upon early discussions of logic gate operation (introduced in Book 1) to explore gate speed and advanced gate topologies.

Fundamentals of Electronics has been designed primarily for use in upper division courses in electronics for electrical engineering students and for working professionals. Typically such courses span a full academic year plus an additional semester or quarter. As such, *Oscillators and Advanced Electronics Topics* and the three companion book of *Fundamentals of Electronics* form an appropriate body of material for such courses.

KEYWORDS

oscillators, phase-shift oscillator, Wien-Bridge oscillator, Colpitts oscillator, Hartley oscillator, multivibrator, waveform shaping, 555 timer, silicon controlled rectifier (SCR), Triac, voltage regulator, transient suppression, overvoltage protection, analog to digital conversion, voltage controlled oscillator (VCO), phase-locked look (PLL), filter, modulator, demodulator, TTL, ECL

Contents

Preface

This final book of this work focuses on four significant branches of electronic circuitry that are not necessarily in the direct path taken by the three previous books. Each is particularly significant in its own right. Discussion here provides an introduction to these specialized topics:

- Oscillator discussions focus on both linear and non-linear oscillators.

- Power circuit discussions focus on DC power.

- Communication circuit discussions focus on high-frequency operation and the frequency translation of signals.

- Digital circuit discussions focus on the non-linear operation of digital gates and the speed at which they operate.

The discussion of linear oscillators is based on a foundation derived from feedback amplifier principles. Linear oscillators are shown to be amplifiers driven into a region of oscillatory instability characterized by the Barkhausen criterion. Colpitts, Hartley and Wein-Bridge topologies form the basis for primary oscillator discussion. Stabilization of the frequency of oscillation through the use of a crystal is explored. Non-linear oscillators (or waveform generators) are explored through discussions of multivibrator circuits. Other non-linear circuits, such as pulse generators and Schmitt triggers, are derived from multivibrator characteristics. An introduction to arbitrary waveform generation concludes these two chapters.

Electronic power circuitry is vital in the operation of all electrical apparatus. The focus of discussion on power electronics presented here is directed at providing clean, reliable power to electronic circuitry. Central to the discussion is the design, operation, and use of voltage regulators. Another topic of particular interest to the electronics designer, transient suppression along with overvoltage protection, is presented. The thyristor family of electronic devices is introduced.

Communication circuits typically translate baseband signals to a higher frequency range for transmission. In order to accomplish this task and reverse it at the receiver, a variety of fundamental building blocks are necessary. An introduction to these building blocks is presented here. Voltage-controlled oscillators and mixers form the primary building blocks discussed for transmission modulators, while mixers and phase-lock loops are presented as common in receiver demodulators. A discussion of the trade-offs between active and passive filters complements previous discussion in Chapter 9 (Book 3). Analog-to-digital conversion is introduced.

While an introduction to the basic operation of a few logic gates was presented in Chapters 3 and 4 (Book 1), discussion there focused on transistor operation rather than gate operation.

Here a closer look at bipolar and FET gate operation, including gate speed is taken. Fundamental problems encountered when transistors transition between regions of operation are analyzed. Several design alternatives to reduce these problems are discussed. Comparisons between gate families are made, but not emphasized. A short introduction to gallium arsenide logic is presented.

Thomas F. Schubert, Jr. and Ernest M. Kim
April 2016

CHAPTER 12

Oscillator Circuits

An electronic oscillator is a circuit that produces a periodic output without an input signal. A harmonic (often called linear) oscillator, which is the topic of this chapter, is a subset of electronic oscillators that produce an output signal that is approximately sinusoidal. The oscillation is based on a resonant circuit often designed using inductors and capacitors. Crystals may be used to closely control the oscillation frequency.

Modern applications of oscillators include audio and electronic communication systems. These systems often contain several oscillators including crystal-controlled reference oscillators, voltage-controlled oscillators (VCOs), and voltage-controlled crystal oscillators. Although many integrated circuits exist for generating periodic signals, discrete oscillator designs have significant advantages over many integrated solutions. In many instances, integrated circuit oscillators cannot meet the high-frequency and low noise requirements of communication systems. Discrete oscillators are also used in high-quality audio systems which require high stability and low noise.

The basic feedback amplifier topologies, presented in previous chapters, are used to analyze and design oscillators. These circuits can be used to generate essentially sinusoidal waveforms by carefully designing the amplifier to operate at the critical point where the loop gain, Af, is -1. When the loop gain is at the critical point, the circuit is oscillatory and delivers a sinusoidal waveform without an externally applied input signal.

Harmonic oscillator designs that incorporate feedback amplifier topology with the loop gain at the critical point are used in a variety of electronic systems. These circuits are commonly called linear oscillators since they generate the waveforms through resonance phenomenon: that is, a frequency selective feedback circuit is used to amplify the frequency of interest. An alternate approach using non-linear signal waveshaping electronic circuits is often employed in oscillator circuit design. These nonlinear circuits, called multivibrators, are discussed in Chapter 13.

12.1 LINEAR ANALYSIS

The basic topology of a harmonic oscillator, shown in Figure 12.1, is identical to the feedback amplifier topology introduced in Chapter 8 (Book 2). The open-loop amplifier gain, $A(s)$, and feedback return ratio, $f(s)$, are frequency selective. An input signal, $X_i(s)$, is shown in Figure 12.1 for the sake of analysis: oscillators do not require an input signal to generate a sinusoidal output waveform. In oscillator circuits, $X_o(s)$ is non-zero for $X_i(s) = 0$.

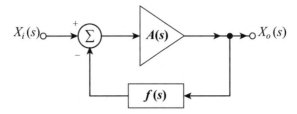

Figure 12.1: Basic topology of a harmonic oscillator with injected input signal.

The closed-loop gain of the circuit shown in Figure 12.1 is,

$$A_f(s) = \frac{A(s)}{1 + A(s) f(s)}, \tag{12.1}$$

where

$$A_f(s) = \frac{X_o(s)}{X_i(s)}.$$

The output signal is therefore,

$$X_o(s) = \frac{A(s)}{1 + A(s) f(s)} X_i(s). \tag{12.2}$$

In electronic oscillators, the circuit is not excited by an external time varying source: that is, the input signal $X_i(j\omega) = 0$. Therefore, for a non-zero output $X_o(s)$, Equation (12.2) must yield a loop gain $A(j\omega_o) f(j\omega_o) = -1$, at the frequency of oscillation, ω_o. From the discussion in Chapter 11 (Book 3), this loop gain condition causes instability resulting in amplifier oscillation.

The condition of unity loop gain with a 180° phase shift is called the *Barkhausen criterion*. In order to achieve the requisite phase shift, reactive elements must be used in the feedback loop. Since the feedback element is reactive, the signal phase shift is invariably a function of frequency. This implies that there is only one frequency where the Barkhausen criterion is satisfied. The circuit oscillates at the frequency where the Barkhausen criterion is satisfied.

When the oscillator is designed for $A(j\omega_o) f(j\omega_o) = -1$, there is a chance that oscillation will cease due to variations in the characteristics of the active elements of the amplifier caused by slight temperature variations, power supply noise, etc., that may force $|A(j\omega) f(j\omega)| < 1$. If the magnitude of the loop gain becomes smaller than unity, the oscillation will decay and then cease. Therefore, an oscillator with the loop gain, $A(j\omega_o) f(j\omega_o)$ exactly equal to -1 is not realizable in practice. If $|A(j\omega) f(j\omega)| > 1$, the signal that is fed back into the mixing point (the summing node) from the feedback network will be smaller than at the input to $f(s)$. Therefore, the output will appear larger than that of the pervious trip around the loop. This larger voltage at the output will then reappear as a still larger voltage on the next trip around the loop. It therefore appears that for $|A(j\omega)f(j\omega)| > 1$, the amplitude of the oscillation increases without limit. In fact, the output signal amplitude is limited by the power supply rails and the onset of nonlinear operation of the active devices of the amplifier. As the oscillations increase in amplitude, the nonlinearity of the circuit becomes more apparent. The onset of nonlinearity limits the amplitude of the oscillation.

In practical oscillators, it is necessary to design $|A(j\omega)f(j\omega)|$ to be slightly greater than unity: To ensure that the magnitude of the loop gain will not fall below unity with electronic noise or variations in transistor and circuit parameters, the loop gain, $|A(j\omega)f(j\omega)| \geq 1.05$. As stated before, the amplitude of oscillation in practical harmonic oscillators is limited by the onset of nonlinearity. In many cases, the excursion into nonlinear regions of operation by oscillators is small, thereby allowing linear theory to be used to design the circuits. Additionally, the reactive feedback network is frequency selective and serves a dual purpose in feedback oscillator design. First, it ensures that the Barkhausen criterion is satisfied and that the conditions of oscillation are met only at the desired frequency which is significantly lower than the high cutoff frequency of the amplifier. Second, the reactive feedback network removes the harmonics of the distorted signal caused by the onset of nonlinearity so that a relatively pure sinusoid at the fundamental oscillation frequency appears at both the output (sampling point) and the mixing point.

12.1.1 DESIGN OF PRACTICAL OSCILLATORS

The design of oscillators is more an art than an exact science since the steady-state operating conditions cannot be accurately predicted by simple mathematical techniques. At the onset of nonlinearity, the linear models used for the analysis of amplifiers are not representative of all aspects of circuit behavior. However, linear analysis is useful in predicting many aspects of oscillator operation.

In oscillators, the output of the circuit must be fed back into the input with a gain slightly greater than unity and with a phase shift of $0°$ (or some multiple of $360°$): this condition of oscillation only occurs at the frequency of the sinusoidal oscillation. If the phase shift through the reactive feedback network is assumed to be independent of the amplifier operating conditions, the steady-state frequency of oscillation will be identical to the initial oscillation frequency in the transient state at the beginning of oscillation. For independent amplifier operation, accurate predictions of the initial oscillation frequency can be made using linear small-signal analysis. Linear small-signal analysis yields the minimum required amplifier gain and operating frequency of an oscillator. In most oscillator designs, the reactive network has an effect on amplifier operation. To counteract the loading effect, the magnitude of the loop gain at the oscillation frequency is $1.3 > |A(j\omega)f(j\omega)| > 1.05$ to insure generation of a sinusoid with minimum harmonic distortion. Higher values of $|A(j\omega)f(j\omega)|$ results in an oscillatory waveform with higher harmonic content. The transient time prior to steady-state oscillation is shorter for higher loop gains at the oscillation frequency.

Another factor that complicates oscillator design is the frequency dependence of the reactive components used in the feedback network. In particular, for radio frequency (RF) applications, a capacitor larger than a few hundred picofarads tends to appear inductive above 10 MHz. An inductor may appear capacitive at higher frequencies due to stray capacitance between the windings of the inductor. These so-called "parasitic" effects are difficult to model using conventional circuit theory. In fact, the parasitic effects of the reactive components may cause the circuit to oscillate at a frequency other than that predicted by small-signal analysis: that is, an oscillator circuit that appears to meet required specifications may oscillate not only at the designed frequency of oscillation, but at a lower frequency (a phenomenon called "motor-boating") and at one or more frequencies higher than the designed oscillating frequency. In most cases, these effects can be mitigated by employing high quality inductors and by connecting small (100 to 300 pF) capacitors in parallel with all bypass and coupling capacitors. These smaller capacitors provide an effective short circuit at frequencies where the larger capacitors become inductive.

The design process of an oscillator begins with the determination of the circuit topology and the value of its elements. The circuit element values determine the desired oscillating frequency. However, such an analysis does not predict circuit power output, efficiency, waveform purity, frequency stability, or temperature and power supply variation sensitivities. These effects are most often resolved in the design by computer simulation and hardware prototyping to adjust circuit component values until the desired overall performance is achieved. In practice, many designers use a few "pet" oscillator circuits that are adapted to fulfill the required performance specifications.

12.1.2 FREQUENCY STABILITY

Frequency stability is expressed in terms of the amount of frequency change with respect to a change in a particular circuit parameter, such as the small-signal transistor current gain h_{fe}. Assume that is one of these circuit parameters, so that a change in $\Delta\kappa$ about its equilibrium point causes a change in the loop gain of a circuit. That is,

$$\Delta(Af) = \frac{\partial(Af)}{\partial\kappa}\Delta\kappa. \tag{12.3}$$

Stable frequency operation requires that the loop gain remain invariant. From Equation (12.1), the condition for stable oscillator operation implies that

$$\frac{\partial(Af)}{\partial\kappa} \to 0.$$

For variations in κ, the loop gain of an oscillator is expressed as,

$$Af = 1 + \Delta(Af). \tag{12.4}$$

This corresponds to a new oscillating frequency and amplitude of oscillation:

$$\omega = \omega_o + \Delta\omega, \tag{12.5a}$$

and

$$v = v_o + \Delta v, \tag{12.5b}$$

where ω_o and v_o are the initial equilibrium oscillating frequency and amplitude, respectively. The difference between the new and old equilibrium points are,

$$\Delta\,(Af) = \frac{\partial\,(Af)}{\partial\kappa}\Delta\kappa + \frac{\partial\,(Af)}{\partial v}\Delta v + \frac{\partial\,(Af)}{G\omega}\Delta\omega = 0. \tag{12.6}$$

Solving Equation (12.6) for $\Delta\omega$,

$$\Delta\omega = \frac{\dfrac{\partial\,(Af)}{\partial\kappa}\Delta\kappa + \dfrac{\partial\,(Af)}{\partial v}\Delta v}{\dfrac{\partial\,(Af)}{\partial\omega}}. \tag{12.7}$$

Three methods for stabilizing oscillators, using Equation (12.7) are:

1. The numerator of Equation (12.7) is set equal to zero. That is,

$$\Delta v = \frac{\dfrac{\partial\,(Af)}{\partial\kappa}}{\dfrac{\partial\,(Af)}{\partial v}}\Delta\kappa. \tag{12.8}$$

   Equation (12.8) states that a rate of change in the circuit parameter κ must be counterbalanced by a change in another circuit parameter.

2. The denominator of Equation (12.7) must approach infinity. In order for $\Delta\omega$ to approach zero, $\partial(Af)/\partial\omega$ must approach infinity. An alternate expression that will allow for the solution to this condition is,

$$\frac{\partial\,(Af)}{\partial\omega} = f\frac{\partial A}{\partial\omega} + A\frac{\partial f}{\partial\omega}. \tag{12.9}$$

   To simplify the analysis, assume that A is independent of ω. Then Equation (12.9) simply becomes,

$$\frac{\partial\,(Af)}{\partial\omega} = A\frac{\partial f}{\partial\omega}. \tag{12.10}$$

   Since the feedback factor f describes a reactive feedback network, it is complex. Therefore, the amplitude and phase relationship must obey the relationship,

$$A\frac{\partial f}{\partial\omega} \to \infty. \tag{12.11}$$

   Therefore, the magnitude and phase of f are: $A\frac{\partial|f|}{\partial\omega} \to \infty$ and $A\frac{\partial\theta_f}{\partial\omega} \to \infty$.
   Assuming that A is independent of ω,

$$\frac{\partial\,(Af)}{\partial\omega} = A\frac{\partial f}{\partial\omega}.$$

Since the slope $\partial|f|/\partial\omega$ is directly proportional to the quality factor (Q) of the oscillator circuit, the fulfillment of the requirement of Equation (12.11) requires that a high quality factor (Q) circuit is required for stability.

3. The open loop gain of the oscillator circuit can be used to stabilize the circuit. In this case, A must be made very large. Therefore, f must be small to meet the oscillation criterion of $Af = -1$. However, the derivative of f with respect to ω must be large to fulfill the requirement of Equation (12.11).

12.2 *RC* OSCILLATORS

Simple *RC* oscillators are commonly used in audio frequency applications that span the frequency range from several hertz to several tens of kilohertz. The two most commonly used oscillator circuits are the *RC* phase-shift and Wien-bridge oscillators.

12.2.1 PHASE-SHIFT OSCILLATOR

The phase-shift oscillator is one of the simplest oscillators to design and construct in the audio frequency range. The oscillator exemplifies the simple principles and conditions of oscillation discussed in Section 12.1. A simple OpAmp-based phase-shift oscillator is shown in Figure 12.2.

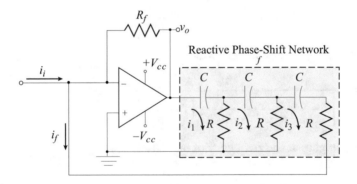

Figure 12.2: OpAmp-based phase-shift oscillator.

In this circuit, an inverting OpAmp amplifier is followed by an *RC* "ladder" network consisting of three cascaded arrangements of a resistor R and capacitor C. The three resistors and three capacitors in the feedback network have identical values. The output of the reactive phase-shift ladder network is returned to the input of the inverting OpAmp amplifier. This feedback configuration is the shunt-shunt topology. Therefore, the analysis used for shunt-shunt feedback topologies is used. The amplifier gain is,

$$A = R_M = -R_f. \tag{12.12}$$

The feedback ratio is,

$$f = \frac{i_f}{v_o} = -\frac{i_3}{v_o}. \tag{12.13}$$

Using the Barkhausen criterion,

$$Af = -1 = -R_f \left(-\frac{i_3}{v_o} \right) = \frac{R_f i_3}{v_o}. \tag{12.14}$$

If the loading of the reactive phase-shift network on the inverting amplifier can be neglected, the output of the inverting amplifier shifts the input signal by 180°. At a particular frequency, the *RC* network shifts the phase by an additional 180°, resulting in a total phase shift of 0° from the output of the OpAmp amplifier to the input. The circuit oscillates at this frequency, provided that the amplifier gain is sufficiently large.

In order to solve for the loop gain, mesh equations for Mesh 1, Mesh 2, and Mesh 3 are formulated:

$$\text{Mesh 1: } v_o - \frac{i_1}{j\omega C} - (i_1 - i_2) R = 0$$

$$\text{Mesh 2: } (i_1 - i_2) R - \frac{i_2}{j\omega C} - (i_2 - i_3) R = 0 \tag{12.15}$$

$$\text{Mesh 3: } (i_2 - i_3) R + \frac{i_3}{j\omega C} - i_3 R = 0.$$

Solve for the current i_3 using Cramer's Rule,

$$i_3 = \frac{v_o}{R \left[1 - \dfrac{5}{\omega^2 R^2 C^2} + \dfrac{1}{j\omega RC} \left(6 - \dfrac{1}{\omega^2 R^2 C^2} \right) \right]}. \tag{12.16}$$

Therefore, the loop gain at the frequency of oscillation is,

$$fA = \frac{-i_3}{v_o} R_f = \frac{R_f}{R \left[1 - \dfrac{5}{\omega^2 R^2 C^2} + \dfrac{1}{j\omega RC} \left(6 - \dfrac{1}{\omega^2 R^2 C^2} \right) \right]} = -1. \tag{12.17}$$

For equality to occur, the denominator must be real and the imaginary part zero

$$6 - \frac{1}{\omega^2 R^2 C^2} = 0. \tag{12.18}$$

The frequency of oscillation is found by solving for ω,

$$\omega RC = \sqrt{6} \quad \Rightarrow \quad \omega = \frac{1}{RC\sqrt{6}}. \tag{12.19}$$

At that frequency of oscillation, the feedback factor $f = -1/29R$. For proper oscillation, the magnitude of the loop gain must be slightly greater than unity

$$fA = \left| -R_f \left(\frac{-1}{29R} \right) \right| > 1 \quad \Rightarrow \quad \frac{R_f}{R} > 29. \tag{12.20}$$

This condition states that an inverting amplifier must have a voltage gain. $|A| > 29$: the resistors, R_f and R, are related by the required gain of the amplifier.

Example 12.1   Design of an OpAmp phase-shift oscillator

Design an OpAmp phase-shift oscillator to produce a 1 kHz ± 10% sinusoidal wave. Use a μA741 OpAmp and power supply voltages, ±15 V. Verify the design using SPICE.

Solution:

The circuit topology of the oscillator is identical to Figure 12.2. Resistance values in the order of 10 kΩ are most practical in OpAmp circuits. The components for the feedback network can be determined using Equation (12.20):

$$\omega = 2\pi \, (1k) = \frac{1}{RC\sqrt{6}}.$$

Since there are more standard resistor value choices available to the designer than capacitor values, choose a reasonable and common capacitor value to initiate the design. For this design, a capacitor value of 0.01 μF is selected:

$$C = 0.01 \, \mu\text{F}.$$

Using the chosen capacitor value, solve for the resistor in the RC ladder network:

$$R = \frac{1}{2\pi \, (1\,\text{k}) \, (0.01 \times 10^{-6}) \, \sqrt{6}} = 6.497 \, \text{k} \approx 6.49 \, \text{k}\Omega.$$

The magnitude of the gain of the inverting OpAmp amplifier at midband must be greater than 29 by at least 5%. Choose a gain of $1.2(29) = 34.8$. The feedback resistor R_f of the inverting amplifier is then:

$$R_f = - (A_{vo} R) = - [-34.8 \, (6.49\,\text{k})] = 225.9 \, \text{k}\Omega \approx 226 \, \text{k}\Omega.$$

The complete circuit design of the phase-shift oscillator is shown below.

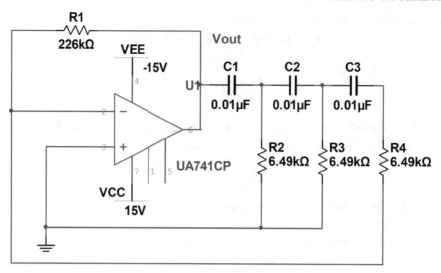

The output of the oscillator is as shown. Note the transient time prior to steady-state oscillation. The transient time is approximately 35 ms. A larger loop gain will shorten the transient time but will also increase harmonic distortion of the steady-state sinusoid.

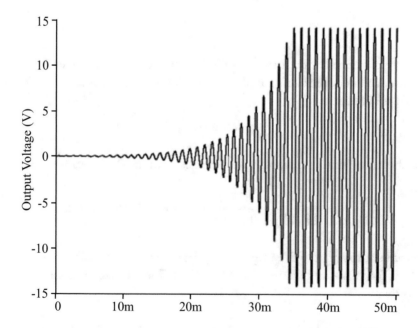

In the accompanying figure, the oscillation frequency is shown to be 935 Hz which meets the design requirement.

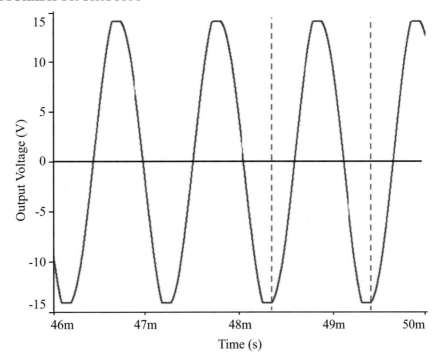

Bipolar Junction Transistor Realization of the Phase-Shift Oscillator

Discrete phase-shift oscillators can be designed using a single BJT or FET amplifier. Figure 12.3 shows a BJT-based phase-shift oscillator. The design procedure for the BJT-based phase-shift oscillator is identical to that of the OpAmp oscillator where the frequency of oscillation is determined by the R and C values of the reactive ladder network according to Equation (12.20). One difference lies in the value of the resistor R' in the RC ladder network that connects the reactive feedback network to the base of the BJT.

The value of R' is adjusted so that,

$$R = R' + (R_B//h_{ie}),\qquad(12.21)$$

where

$$R_B = R_{B1}//R_{B2}.$$

It is evident that the RC phase-shift network is not independent of the BJT amplifier. Therefore, a complete analysis of the circuit is required. Assuming operation in the midband frequency range, the small-signal behavior of the BJT-based phase-shift oscillator in Figure 12.3 can be analyzed using the small-signal equivalent model in Figure 12.4. Assume that h_{oe}^{-1} is very large and can be ignored.

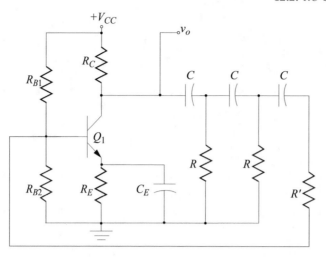

Figure 12.3: BJT-based phase-shift oscillator.

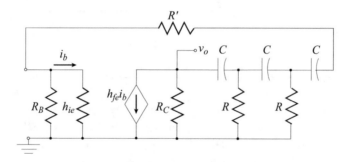

Figure 12.4: Small-signal equivalent circuit of a BJT-based phase-shift oscillator.

The small-signal equivalent circuit of the BJT-based phase-shift oscillator shown in Figure 12.4 is modified in Figure 12.5 to allow calculation of the loop gain of the circuit. The equivalent resistance R_i is,

$$R_i = R_B // h_{ie}. \qquad (12.22)$$

The loop gain for the circuit in Figure 12.5 is most conveniently represented in terms of currents,

$$Af = \frac{i_3}{i_b}. \qquad (12.23)$$

The loop gain is found by using the method of analysis shown in the phase-shift oscillator circuit. When the oscillator is designed so that $R_B \gg h_{ie}$ (that is, $R_i = R_B // h_{ie} \approx h_{ie}$), Equa-

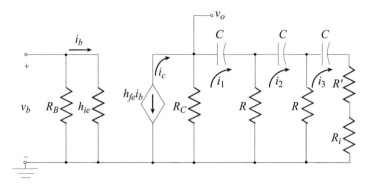

Figure 12.5: Equivalent circuit used to calculate loop gain.

tion (12.23) yields,

$$A(j\omega) f(j\omega) = \frac{h_{fe}}{3 + \dfrac{R}{R_C} - j\dfrac{4}{RX} - j\dfrac{6}{R_C X} - \dfrac{1}{R_C^2 X^2} - \dfrac{5}{R R_C X^2} + j\dfrac{1}{R^2 R_C X^3}}, \quad (12.24)$$

where $X = \frac{1}{\omega C}$.

The Barkhausen criterion for oscillation requires that $A(j\omega) f(j\omega) = 1 \angle 180° = -1$. Therefore, oscillation occurs when the imaginary part of the denominator in Equation (12.24) is zero,

$$0 = j\left(\frac{4}{RX} - \frac{6}{R_C X} + \frac{1}{R^2 R_C X^3}\right)$$
$$= j\left[\frac{1}{\omega}\left(\frac{4}{RC} - \frac{6}{R_C C}\right) + \frac{1}{\omega^3 R^2 R_C C^3}\right]. \quad (12.25)$$

Equation (12.26) simplifies to,

$$\frac{1}{\omega^2 R R_C C^2} = 4 + \frac{6R}{R_C}. \quad (12.26)$$

Solving for the oscillation frequency, ω, in Equation (12.26) results in the expression:

$$\omega = \frac{1}{RC\sqrt{\dfrac{4R_C}{R} + 6}}. \quad (12.27)$$

The real part of the loop gain $A(j\omega) f(j\omega) = -1$. Under this condition, the imaginary component of the denominator of Equation (12.24) is zero. Therefore, at the oscillation condition,

$$-1 = \frac{h_{fe}}{3 + \dfrac{R}{R_C} - \dfrac{1}{\omega^2}\left(\dfrac{1}{R_C^2 C^2} + \dfrac{5}{R R_C C^2}\right)}. \quad (12.28)$$

The required current gain, h_{fe}, for the fulfillment of the Barkhausen criterion is found by substituting Equation (12.26) into (12.28) and solving for h_{fe},

$$h_{fe} = 23 + \frac{29}{\frac{R_C}{R}} + 4\frac{R_C}{R}. \tag{12.29}$$

To find the resistor ratio R_C/R for the minimum h_{fe} required to fulfill the Barkhausen criterion, the derivative of h_{fe} with respect to R_C/R is found and set to zero,

$$0 = \frac{d\,h_{fe}}{d\left(\frac{R_C}{R}\right)} = \frac{d}{d\left(\frac{R_C}{R}\right)}\left(23 + \frac{29}{\frac{R_C}{R}} + 4\frac{R_C}{R}\right). \tag{12.30}$$

Equation (12.30) yields the resistor ratio R_C/R,

$$\frac{R_C}{R} = \sqrt{\frac{29}{4}} = 2.7. \tag{12.31}$$

Therefore, to fulfill the oscillation condition, the BJT current gain must be,

$$h_{fe} \geq 23 + \frac{29}{2.7} + 4\,(2.7) = 44.5. \tag{12.32}$$

The result of Equation (12.32) is that the circuit meets the Barkhausen criterion and will oscillate if $h_{fe} \geq 44.5$. Most small-signal BJTs used in oscillator design have current gains well in excess of 44.5. To insure oscillation, the loop gain is adjusted to be slightly greater than unity by adjusting the collector resistor R_C.

MOSFET Realization of the Phase-Shift Oscillator
A depletion NMOSFET-based phase-shift oscillator, shown in Figure 12.6, is simpler to analyze than BJT-based oscillators since the gate (input) resistance is very high. Therefore, the analysis of the FET phase-shift oscillator is similar to the OpAmp oscillator.

   When the FET-based phase-shift oscillator is design so that $R \gg R_D//r_d$, then the loading of the reactive feedback network on the amplifier can be neglected. Similarly, R_g must have much larger value than R to prevent loading of the reactive feedback network. The equation for the loop gain is found using the methods described for the BJT-based and OpAmp-based phase-shift oscillators. The loop gain of the NMOSFET-based phase-shift oscillator in Figure 12.6 is,

$$A\,(j\omega)\,f\,(j\omega) = \frac{j\,g_m\,(R_D//r_d)\,\omega^3 R^3 C^3}{(1 - 6\omega^2 R^2 C^2) + j\omega RC\,(5 - \omega^2 R^2 C^2)}, \tag{12.33}$$

where g_m is the mutual conductance of the FET.

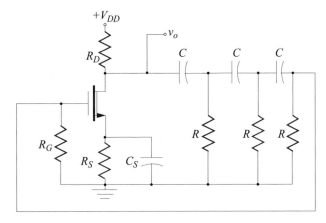

Figure 12.6: Depletion NMOSFET-based phase-shift oscillator; bias resistor $R_G \gg R$.

At the oscillation condition,

$$\omega_o = \frac{1}{RC\sqrt{6}} \quad \Rightarrow \quad f_o = \frac{1}{2\pi RC\sqrt{6}}. \tag{12.34}$$

Substituting Equation (12.34) into (12.33) yields the expression for the loop gain at the oscillation frequency,

$$A(j\omega) f(j\omega) = \frac{g_m (R_D // r_d)}{29}. \tag{12.35}$$

From Equation (12.35), the gain of the FET amplifier must be $g_m R_L = -29$. This condition implies that the magnitude of the gain of the FET amplifier must be at least 29. This conclusion is identical to that of the OpAmp phase-shift oscillator.

12.2.2 WIEN-BRIDGE OSCILLATOR

The Wien-bridge oscillator circuit uses a balanced bridge as its reactive feedback network. The OpAmp implementation of the Wien-bridge oscillator is shown in Figure 12.7 where the bridge network is clearly depicted. When using an OpAmp as its gain element, the oscillator is assumed to operate at a frequency significantly lower than the unity gain frequency of the OpAmp.

The feedback configuration of the Wien-bridge oscillator is the series-shunt topology. To simplify the analysis of the circuit, Figures 12.8a and b are used. To find the loop gain of the oscillator circuit, the feedback path at the output of the OpAmp is broken and an external voltage v_L is applied as shown in Figure 12.8b. Auxiliary voltages v_i, v_1, and v_2 are indicated to simplify the analysis.

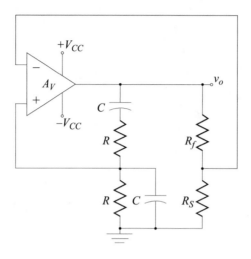

Figure 12.7: OpAmp Wien-bridge oscillator circuit.

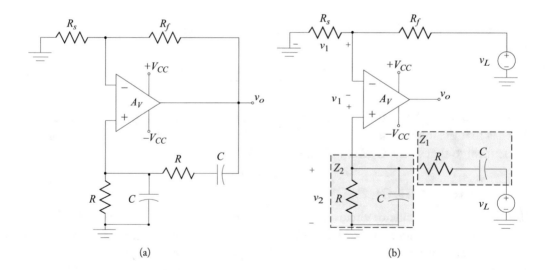

Figure 12.8: (a) OpAmp Wien-bridge oscillator (re-drawn); (b) Circuit used to calculate loop gain.

The open-loop gain of the OpAmp, A_V (very large positive gain) and the output voltage is $v_o = A_V v_i$. The loop gain is therefore,

$$Af = -\frac{v_o}{v_L} = -\frac{v_i}{v_L} A_V. \tag{12.36}$$

Noting that $v_i = v_2 - v_1$ and $A = A_V$, the feedback factor is,

$$f = -\frac{v_i}{v_L} = -\left(\frac{v_2 - v_1}{v_L}\right) = -\left(\frac{Z_2}{Z_1 + Z_2} - \frac{R_S}{R_S + R_f}\right). \tag{12.37}$$

Therefore, the loop gain for the circuit is,

$$Af = -\frac{v_o}{v_L} = -A_V \left[\frac{Z_2}{Z_1 + Z_2} - \frac{R_S}{R_S + R_f}\right]$$

$$= -A_V \left[\frac{\dfrac{R}{(1 + j\omega RC)}}{\dfrac{R}{(1 + j\omega RC)} + R - j\dfrac{1}{\omega C}} - \frac{R_S}{R_S + R_f}\right] \tag{12.38}$$

$$= -A_V \left[\frac{\dfrac{j\omega}{\omega_o}}{\dfrac{j\omega}{\omega_o} + \left(1 + \dfrac{j\omega}{\omega_o}\right)^2} - \frac{R_S}{R_S + R_f}\right],$$

where $\omega_o = \frac{1}{RC}$.

Using the condition for oscillation, $Af = -1$, Equation (12.38) yields,

$$0 = 1 - \left(\frac{\omega}{\omega_o}\right)^2 + j\frac{\omega}{\omega_o}\left[3 - \frac{A_V (R_S + R_f)}{R_S (1 + A_V) + R_f}\right]. \tag{12.39}$$

Equating the real and imaginary parts to zero, Equation (12.39) yields,

$$\omega = \omega_o = \frac{1}{RC}, \tag{12.40}$$

for the oscillation frequency, and

$$\frac{R_f}{R_S} = \frac{2A_V + 3}{A_V - 3} \approx 2, \tag{12.41}$$

for the ratios of the gain resistors R_f and R_S.

12.3 *LC* OSCILLATORS

LC oscillators are commonly used in radio frequency applications that span the frequency range from several hundred kilohertz to several hundred megahertz. The two most commonly used oscillator circuits are the Colpitts and Hartley oscillators. These oscillators are configured in the general form shown in Figure 12.9. The amplifier may be an OpAmp, BJT, or FET amplifier. The amplifier gain is A_v, the input resistance to the amplifier is R_i, output resistance is R_o, and complex impedances Z_1, Z_2, and Z_3. The feedback topology of the circuit in Figure 12.9 is a series-shunt feedback.

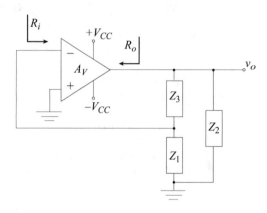

Figure 12.9: General form of many oscillator circuits.

The load impedance of the circuit is,

$$Z_L = [(Z_1 \, // \, R_i) + Z_3] \, // Z_2. \tag{12.42}$$

The open loop gain is,

$$A = \frac{-A_v \, Z_L}{Z_L + R_o}. \tag{12.43}$$

The feedback factor is,

$$f = \frac{Z_1 \, // \, R_i}{(Z_1 \, // \, R_i) + Z_3}. \tag{12.44}$$

The loop gain is found by multiplying Equations (12.43) and (12.44),

$$Af = \frac{-A_v \, (Z_1 \, // \, R_i) \, Z_2}{R_o \, [(Z_1 \, // \, R_i) + Z_2 + Z_3] + Z_2 \, [(Z_1 \, // \, R_i) + Z_3]}. \tag{12.45}$$

If the impedances are purely reactive (inductive or capacitive), then $Z_1 = jX_1, Z_2 = jX_2, Z_3 = jX_3$, where $X = \omega L$ for inductors and $X = -1/\omega C$ for capacitors, the loop gain is,

$$Af = \frac{A_v \left(\dfrac{X_1^2 R_i + j X_1 R_i^2}{R_i^2 + X_1^2} \right) X_2}{j R_o \left[\left(\dfrac{X_1 R_i^2}{R_i^2 + X_1^2} \right) + X_2 + X_3 \right] - X_2 \left[\left(\dfrac{X_1^2 R_i}{R_i^2 + X_1^2} \right) + X_3 \right]}. \tag{12.46}$$

For the loop gain to be real with no phase shift and $R_i \gg X_1$, the imaginary component of Equation (12.46) is set equal to zero,

$$0 = X_1 + X_2 + X_3, \tag{12.47}$$

therefore,

$$Af = \frac{A_v X_1 X_2}{X_2 (X_1 + X_3)} = \frac{A_v X_1}{(X_1 + X_3)}. \tag{12.48}$$

From Equation (12.47), it can be seen that the circuit oscillates at a resonant frequency corresponding to the series combination of X_1, X_2, and X_3.

Equation (12.48) is simplified by using the relationship in Equation (12.47),

$$Af = -\frac{A_v X_1}{X_2}. \tag{12.49}$$

For oscillation, the loop gain, $|Af|$, must have a magnitude of at least unity, and X_1 must have the same sign as X_2; that is, they must be the same type of reactance, either both capacitive or inductive. This implies that X_3 must be inductive if X_1 and X_2 are capacitive, or vice versa.

A Colpitts oscillator is a circuit where Z_1 and Z_2 are capacitors, and Z_3 is an inductor. A Hartley oscillator is a circuit where Z_1 and Z_2 are inductors, and Z_3 is a capacitor. In the latter case, mutual inductance between Z_1 and Z_2 will alter the relationships derived above.

LC oscillators are commonly used in Radio Frequency (RF) applications in the frequency range between 100 kHz and several hundred MHz. In this range of frequencies, the simple audio frequency small signal models of active devices (BJTs and FETs) cannot be used. High frequency transistor models must be used in the design and analysis. LC oscillators are found extensively in communication electronics. For example, in AM and FM receivers, station tuning is accomplished by varying a capacitance or inductance.

12.3.1 COLPITTS OSCILLATOR

As stated above, the Colpitts oscillator is a special case of the LC oscillator described in the previous section. A BJT-based Colpitts oscillator circuit is shown in Figure 12.10. In this circuit, the capacitors C_B and C_E are very large coupling and bypass capacitors, respectively.

The small-signal ac model at the output of the BJT-based Colpitts oscillator is shown in Figure 12.11. The equivalent device output capacitance C_o is not included in the equivalent circuit since the high frequency response of BJT amplifiers is limited by its input capacitance, $C_i = C_\pi + \{1 + g_m(R_i//R_C)\}C_\mu$. The input resistance of the BJT amplifier, R_i, is given by: $R_i = (r_b + r_\pi)//R_B$, and the output resistance of the BJT, r_o, is assumed to be very large compared to R_C and R_i.

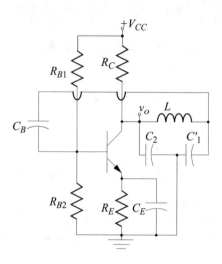

Figure 12.10: BJT-based Colpitts oscillator.

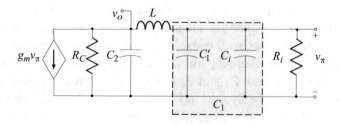

Figure 12.11: Output equivalent circuit for the Colpitts oscillator.

The Colpitts oscillator reactances are identified as:

$$X_1 = \frac{-1}{\omega C_1}, \quad X_2 = \frac{-1}{\omega C_2} \quad \text{and} \quad X_3 = \omega L, \tag{12.50}$$

where

$$C_1 = C_1' + C_i. \tag{12.51}$$

From Equation (12.47), it can be seen that this circuit oscillates at:

$$X_1 + X_2 + X_3 = 0 \quad \Rightarrow \quad -\frac{1}{\omega_o C_1} - \frac{1}{\omega_o C_2} + \omega_o L = 0.$$

Equivalently

$$\omega_o = \sqrt{\frac{C_1 + C_2}{L C_1 C_2}}. \tag{12.52}$$

The requirements for proper oscillation come from the Barkhausen criterion applied to Equation (12.49):

$$|Af| = \left| A_v \frac{X_1}{X_2} \right| = \left| A_v \frac{C_2}{C_1} \right| > 1. \tag{12.53}$$

Where A_v is the midband voltage gain (obtained from Figure 12.11 with capacitors replaced by open circuits and inductors replaced by short circuits).

$$A_v = \frac{v_o}{v_\pi} = -g_m \left(R_C // R_i \right) \approx -g_m \left(R_C // r_\pi \right) = \frac{-\beta_F R_C}{R_C + r_\pi}. \tag{12.54}$$

Substitution of Equation (12.54) into (12.53) yields the Colpitts Barkhausen criterion:

$$\frac{\beta_F R_C}{R_C + r_\pi} > \frac{C_1}{C_2}. \tag{12.55}$$

Example 12.2 BJT-based Colpitts Oscillator

Complete the design of the circuit shown for an oscillation frequency of 10.7 MHz. Assume that the transistor parameters are:

$$\beta_F = 200,$$
$$V_A = 150\,\text{V},$$
$$r_b = 30\,\Omega,$$
$$C_\mu = 3\,\text{pF},$$
$$f_T = 250\,\text{MHz}.$$

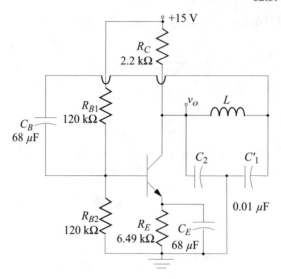

Solution:

First determine the bias condition of the oscillator.

Applying KVL to the base-emitter loop,

$$0 = \frac{V_{CC}R_{B2}}{R_{B1} + R_{B2}} - \frac{I_C}{\beta_F}(R_{B1} // R_{B2}) - V_\gamma - I_C \frac{\beta_F + 1}{\beta_F} R_E.$$

Substituting in the appropriate values yields the collector current,

$$I_C = 1 \, \text{mA}.$$

Checking the collector-emitter loop results in $V_{CE} = 6.3 \, \text{V}$.

The hybrid-π parameters of interest are:

$$g_m = \frac{|I_C|}{V_t} = \frac{0.001}{0.026} = 38.5 \, \text{mS} \qquad r_\pi = \frac{\beta_F}{g_m} = \frac{200}{0.0385} = 5.2 \, \text{k}\Omega.$$

The output resistance of the BJT, r_o, is significantly large enough to ignore in this design. Also, $R_B \gg r_\pi$.

The capacitance C_π is,

$$C_\pi = \frac{g_m}{\omega_T} - C_\mu = \frac{0.0385}{2\pi \left(250 \times 10^6\right)} - 3 \times 10^{-12} = 21.5 \, \text{pF}.$$

The Miller's equivalent input capacitance of the BJT is,

$$C_i = C_\pi + C_\mu \left(1 + g_m R\right) = C_\pi + C_\mu \left\{1 + g_m \left[R_{B1} // R_{B2} // R_C // (r_b + r_\pi)\right]\right\}$$
$$= \left[21.5 + 3\left[1 + 0.0385\left[120 \times 10^3 // 120 \times 10^3 // 2.2 \times 10^3 // \left(30 + 5.2 \times 10^3\right)\right]\right]\right] \times 10^{-12}$$
$$= 199 \, \text{pF}.$$

Then

$$C_1 = C_1' + C_i = 0.01\,\mu\text{F} + 199\,\text{pF} = 10.2\,\text{nF}.$$

Applying the Colpitts Barkhausen criterion using a factor of 1.25, one can determine the value of C_2

$$\frac{\beta_F R_C}{(R_C + r_\pi)} = 1.25 \frac{C_1}{C_2}$$

$$C_2 = \frac{1.25\,(R_C + r_\pi)\,C_1}{\beta_F R_C} = \frac{1.25\,[2200 + 5200]\,(10.2 \times 10^{-9})}{200\,(2200)} = 214\,\text{pF}.$$

Select $C_2 = 220\,\text{pF}$.

The desired frequency of oscillation is 10.7 MHz as expressed as

$$f_o = \frac{1}{2\pi} \sqrt{\frac{C_1 + C_2}{LC_1C_2}}.$$

Solving for L

$$L = \frac{C_1 + C_2}{(2\pi f_o)^2 C_1 C_2} = \frac{(10.2 \times 10^{-9} + 220 \times 10^{-12})}{\left[2\pi\,(10.7 \times 10^6)\right]^2 (10.2 \times 10^{-9})\,(220 \times 10^{-12})}$$

$$= 1.03\,\mu\text{H} \approx 1.0\,\mu\text{H}.$$

SPICE Results:

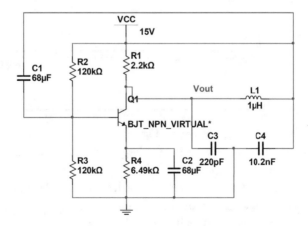

The resulting output at the collector of the BJT is shown below. Note the finite transient time to steady-state oscillation.

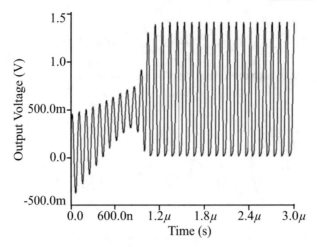

The steady-state output of the oscillator is shown below. The period of oscillation is 98.3 ns which correspond to 10.2 MHz. The oscillation frequency of the simulated circuit is within 5% of the desired oscillating frequency of 10.7 MHz.

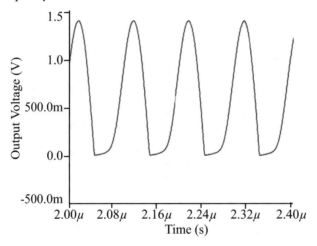

A FET-based Colpitts oscillator, shown in Figure 12.12, is somewhat simpler to design and analyze since the input (gate) resistance is very large. The capacitors C_G and C_S are very large coupling and bypass capacitors, respectively.

If the circuit is designed with a large value of R_G, the small-signal model at the output of the FET-based Colpitts oscillator is shown in Figure 12.11.

The equivalent device output capacitance C_o is not included in Figure 12.13 since the high frequency response of FET amplifiers is limited by $C_i = C_{gs} + [1 + g_m (R_D//r_d)] C_{gd}$. Once again $C_1 = C_1' + C_i$.

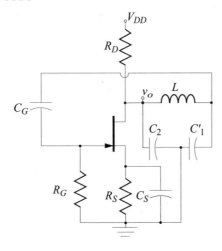

Figure 12.12: JFET-based Colpitts oscillator.

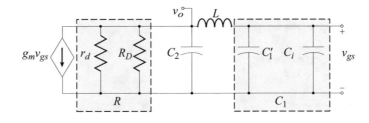

Figure 12.13: Output equivalent circuit for the Colpitts oscillator in Figure 12.12.

As was the case for the BJT Colpitts oscillator, the circuit will oscillate at

$$\omega_o = \sqrt{\frac{C_1 + C_2}{LC_1C_2}}, \tag{12.56}$$

and the requirements for proper oscillation come from the Barkhausen criterion applied to Equation (12.49):

$$|Af| = \left| A_v \frac{X_1}{X_2} \right| = \left| A_v \frac{C_2}{C_1} \right| > 1. \tag{12.57}$$

Where A_v is the midband voltage gain (obtained from Figure 12.11 with capacitors replaced by open circuits and inductors replaced by short circuits).

$$A_v = \frac{v_o}{v_\pi} = -g_m \left(R_D // r_d \right). \tag{12.58}$$

Substitution of Equation (12.54) into (12.53) yields the Colpitts Barkhausen criterion:

$$g_m \left(R_D // r_d \right) > \frac{C_1}{C_2}.$$

(12.59)

An OpAmp-based Colpitts oscillator is shown in Figure 12.14. The small output resistance of the OpAmp is included in the analysis.

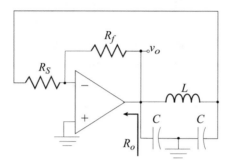

Figure 12.14: OpAmp-based Colpitts oscillator.

The oscillation condition occurs at

$$\omega_o = \sqrt{\frac{2}{LC}}.$$

(12.60)

The oscillator circuit should be designed so that,

$$\frac{R_f}{R_S} > 1.$$

(12.61)

The advantage of using Colpitts oscillators comes from the fact that tuning a single inductor allows for variation in the oscillation frequency. This is especially advantageous since inductors can easily be tuned by introducing ferrite material in through the cross-section of the inductor coil.

12.3.2 HARTLEY OSCILLATOR

The Hartley oscillator circuit is similar to the Colpitts oscillator with one major difference: the capacitors and inductors have been "swapped." A Colpitts oscillator can be mapped into a Hartley oscillator by exchanging jX_L and $-jX_C$. A FET-based Hartley oscillator circuit is shown in Figure 12.15, where C_D is a large coupling capacitor required to isolate the DC and AC models. The inductors in the circuit are identical and, for simplicity, there it is assumed that there is no mutual inductance coupling between the inductors.

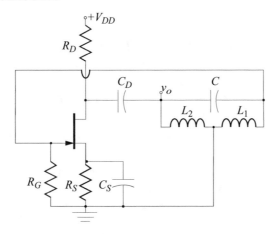

Figure 12.15: FET-based Hartley oscillator.

The Harley oscillator reactances are identified as:

$$X_1 = \omega L_1, \quad X_2 = \omega L_2 \quad \text{and} \quad X_3 = \frac{-1}{\omega C}. \tag{12.62}$$

The midband voltage gain is given by:

$$A_v = \frac{v_o}{v_\pi} = -g_m \left(R_D // r_d \right). \tag{12.63}$$

Oscillation occurs at

$$X_1 + X_2 + X_3 = 0 \quad \Rightarrow \quad \omega_o L_1 + \omega_o L_2 - \frac{1}{\omega_o C} = 0. \tag{12.64}$$

Equivalently

$$\omega_o = \frac{1}{\sqrt{(L_1 + L_2) C}}. \tag{12.65}$$

The condition for oscillation indicates that the circuit will oscillate if

$$\left| g_m \left(R_D // r_d \right) \left(\frac{\omega_o L_1}{\omega_o L_2} \right) \right| \quad \Rightarrow \quad \left| g_m \left(R_D // r_d \right) \right| > \frac{L_2}{L_1}. \tag{12.66}$$

The advantage of using Hartley oscillators comes from the fact that tuning a single capacitor allows for variation in the oscillation frequency. This is especially advantageous since a single voltage variable capacitor called a varactor diode can be used to easily tune the circuit.

12.4   CRYSTAL OSCILLATORS

Crystals are three-dimensional, mechanical oscillating components that oscillate in many differ-ent modes. The crystal oscillations are governed by the crystal piezoelectric properties, and the arrangement and shape of the electrodes attached to the crystal. Crystals are fabricated so that several oscillating and harmonic modes can be used in the design of a circuit. Crystals are available in a wide range of discrete frequencies.

For stable frequency operation, the oscillator should be designed so that a crystal is the controlling element for the oscillation. The crystal oscillator is often a critical component in communications systems and in digital signal processing applications. The circuit symbol for the piezoelectric crystal is shown in Figure 12.16a. The electrical equivalent circuit representing the resonant nature of the piezoelectric crystal is shown in Figure 12.16b. The equivalent circuit is similar to the familiar RLC passive resonant circuit.

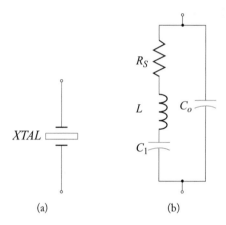

(a)   (b)

Figure 12.16: (a) Circuit symbol for a crystal; (b) Equivalent circuit for a crystal.

The capacitance C_p is in the order of $10\,\mathrm{pF}$ and includes the capacitance associated with the mechanical package and C_s is in the order of $0.05\,\mathrm{pF}$. The crystal equivalent inductance, L, is very large for quartz crystals, on the order of several tens of Henrys. The internal losses of the crystal are represented by R_s which is typically small. The resistive loss, R_s, is related to the quality factor Q of the crystal, since the energy lost during any periodic signal is associated with the dissipative resistance. Q is defined as the ratio of maximum energy stored to the amount lost per periodic cycle. It also determines the bandwidth of resonant circuits. The bandwidth is calculated from Q and the resonant frequency by,

$$BW = \frac{f_o}{Q}. \tag{12.67}$$

The quality factor is related to the dissipative resistance and inductance through the relationship,

$$Q = \frac{\omega_o L}{R_s}. \tag{12.68}$$

Crystals are inherently high Q devices. When used in oscillator circuits, a crystal increases the oscillation stability. This was shown in Section 12.1 where,

$$\frac{\partial (Af)}{\partial \omega} = A \frac{\partial f}{\partial \omega}.$$

Since the slope $\partial |f|/\partial \omega$ is directly proportional to the quality factor (Q) of the oscillator circuit, the high Q characteristic of crystals allows the fulfillment of the requirement of Equation (12.10) that $A(\partial |f|/\partial \omega) \to \infty$. Therefore, crystal oscillators are inherently very stable.

Figure 12.16b provides a simplified equivalent circuit of a crystal at one oscillating frequency. In reality, the crystal has many oscillatory modes (and frequencies). Therefore, a more accurate model of a crystal depicting its many oscillatory frequencies and harmonics is shown in Figure 12.17.

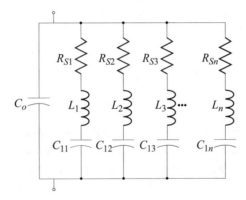

Figure 12.17: Electric circuit equivalent of a crystal showing many oscillatory states.

The circuit in Figure 12.17 contains several series resonant circuits whose frequencies are nearly the odd harmonics of the fundamental oscillatory frequency. The higher resonant frequencies are called overtones of the fundamental. Typical circuit parameters for fundamental, third, and fifth overtone crystals are given in Table 12.1.

If $R_s = 0$, crystal input impedance in a narrow frequency range around f_o is,

$$Z(j\omega) = \frac{(j\omega C_o)^{-1} \left[j\omega L + (j\omega C_1)^{-1} \right]}{j\omega L + (j\omega C_1)^{-1} + (j\omega C_o)^{-1}}. \tag{12.69}$$

Table 12.1: Typical crystal data

f, MHz	Oscillation Mode	R_S, Ω	C_0, pF	C_1, fF	Q
1.0	Fundamental	250	4.0	9.0	65,000
2.0	Fundamental	70	3.5	10.0	110,000
5.0	Fundamental	15	6.0	24.0	86,000
10.0	Fundamental	12	6.0	24.0	50,000
20.0	Fundamental	12	6.0	24.0	25,000
45.0	3rd Overtone	25	5.0	1.5	90,000
100.0	5th Overtone	40	5.0	0.3	130,000

The impedance $Z(j\omega)$ will be zero when L and C_1 are in resonance: that is,

$$f_s = \frac{1}{2\pi\sqrt{LC_1}}.$$ (12.70)

where f_s is the series resonant frequency of the crystal. Conversely, the crystal will have infinite input impedance at the frequency,

$$f_a = \frac{1}{2\pi\sqrt{L\left(\dfrac{C_o C_1}{C_o + C_1}\right)}},$$ (12.71)

where f_a is the antiresonant frequency of the crystal. An ideal crystal (one with $R_s = 0$) behaves as both a series resonant and a parallel resonant circuit with infinite Q. Realistically, all crystals have some series resistance, R_s. For non-zero R_s, the input impedance of a crystal is,

$$Z(j\omega) = \frac{(j\omega C_o)^{-1}\left[j\omega L + R_s + (j\omega C_1)^{-1}\right]}{j\omega L + R_s + (j\omega C_1)^{-1} + (j\omega C_o)^{-1}}.$$ (12.72)

In practice, the input impedance of a crystal at resonance with non-zero R_s is the same as one with $R_s = 0$. The effect of R_s is primarily in the reduction of Q.

Crystal oscillators can be implemented in a variety of topologies. For example, a Pierce oscillator, shown in Figure 12.18, is a simply a Colpitts oscillator where the crystal has replaced the inductor in the feedback path. The Pierce oscillator depends on the inductive component of the crystal to provide a feedback of the proper phase. The crystal will oscillate at a frequency between f_a and f_S.

The circuit in Figure 12.18 is designed and analyzed in the same manner as the LC Colpitts oscillator. Feedback is provided through the crystal which is operating near the series-resonant

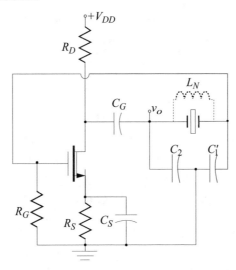

Figure 12.18: Pierce crystal oscillator.

mode frequency. For oscillators operating above 20 MHz, an inductor, L_N, is placed in parallel with the crystal. The purpose of L_N is to neutralize the package capacitance, C_o, where:

$$L_N \approx \frac{1}{(2\pi\ f_o)^2 C_o}. \tag{12.73}$$

The Pierce oscillator is most commonly used with fundamental mode crystals.

A Pierce oscillator can be designed from CMOS inverter used as an amplifier is shown in Figure 12.19. The reactive components of the feedback network consist of C_1, C_2, and *XTAL* (the crystal).

The CMOS amplifier has high-gain inverting amplifier. The reactive feedback network reverses the phase again so that there is positive feedback. This CMOS oscillator is often used as a simple timing circuit.

The effect of adding series or parallel elements to the crystal is quite complex. The resonant frequency is altered, and the dissipative resistance is scaled. Losses due to the added components will also alter the composite Q of the oscillator. To complicate things further, all of the effects mentioned vary as the circuit is tuned or adjusted.

Figure 12.20 shows an improved crystal Colpitts oscillator circuit. The schematic shows a series crystal in the feedback path of the oscillator.

Here the crystal appears as a series resistance R_s (crystal resistance) at the series resonant frequency. Since R_s is small, it can usually be ignored in designing the circuit. The design procedure is identical to an *LC* Colpitts oscillator with an addition of a crystal in the feedback path.

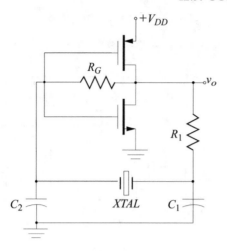

Figure 12.19: CMOS Pierce oscillator.

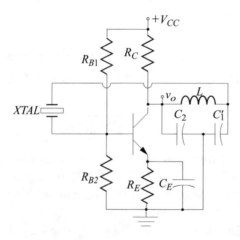

Figure 12.20: Improved crystal Colpitts oscillator.

12.5 CONCLUDING REMARKS

Electronic harmonic oscillator circuits were discussed in this chapter. Although several different configurations were presented, all of the harmonic oscillators shared the Barkhausen criterion for oscillation which simply states that the loop gain of the circuit has unity magnitude gain and a phase of 180°.

Oscillators were designed from BJTs, FETs, and OpAmps. The advantage of using FETs and OpAmps lies in the large input resistance of the devices. Because of their large input resistances, the device does not affect the impedance of the reactive feedback network. However, at higher frequencies, OpAmps cannot be used since the high-frequency response is limited by the relatively low gain-bandwidth product. Therefore, for high frequency RF applications, BJT and FET oscillators are used.

At high frequencies, *LC* reactive networks are commonly used. This is due to the inductive nature of resistors at high frequencies, which prohibits the use of *RC* phase-shift oscillators at these frequencies.

For good frequency stability, crystals are used in the oscillator circuit. Because crystals are inherently reactive, they can be used in place of inductors in oscillator circuits. At higher frequencies, however, a small inductor should be place in parallel with the crystal to counteract the package capacitance.

Summary Design Example: Color Reference Oscillator for Color Television

The color seen on a typical television is emitted from special phosphors. The color cathode ray tube (CRT) has three separate phosphors placed on the screen. When struck by electrons one phosphor emits red, on green, and one blue. The three phosphors are arranged very close to each other in alternating vertical stripes or as tiny points in a matrix pattern. The color that is projected to the user is a combination of the three phosphor emissions.

The electronic signal required to produce color on an otherwise black and white picture signal is called chrominance or *chroma*. The frequency of the chroma signal is 3.5795 MHz. The chroma signal is a vector with the phase angle indicating the hue (tint, or type of color) and magnitude indicating the saturation level or color intensity. A red fire engine is a saturated red object.

There are several oscillator designs that can be considered for the 3.5795 MHz oscillator. They are:

- Phase-shift oscillator

- Wien-bridge oscillator

- Colpitts oscillator

- Hartley oscillator

- Pierce oscillator

Each of the oscillator configurations listed can be design with OpAmps, BJTs, and FETs. OpAmp implementations are not practical in this case since the frequency is higher than most low cost OpAmp gain-bandwidth products. For low cost, it is also common to use inexpensive BJTs for discrete electronic circuits. A common device is the 2N3904 *npn* BJT. The BJT specifications are:

$$\beta_F = 200, \quad C_{ibo} = 8\,\text{pF at } V_{EB} = 0.5\,\text{V} \quad V_A = 150\,\text{V},$$
$$r_b = 30\,\Omega, \quad C_{obo} = 4\,\text{pF at } V_{CB} = 5\,\text{V}, \quad f_T = 300\,\text{MHz at } I_{CT} = 10\,\text{mA}.$$

The oscillator chosen is the BJT Colpitts oscillator. The transistor is to be biased at a modest collector current. Choose $I_C = 1$ mA with a large Thévenin base resistance to simplify analysis. The circuit configuration chosen is shown below.

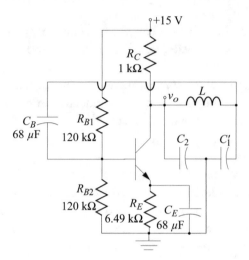

In order to solve for the reactive feedback elements, the effective input and output capacitance of the BJT must be found. The parameters of interest are C_π and C_μ for analytical design.

The hybrid-π parameters of interest are:

$$g_m = \frac{|I_C|}{V_t} = \frac{0.001}{0.026} = 38.5 \, \text{mS} \qquad r_\pi = \frac{\beta_F}{g_m} = \frac{200}{0.0385} = 5.2 \, \text{k}\Omega.$$

The output resistance of the BJT, r_o, is significantly high enough to ignore in the design. Also, $R_B \gg r_\pi 1/1000$.

The capacitance, $C_\mu = C_{obo} = 4 \, \text{pF}$ and the capacitance C_π is,

$$C_\pi = \frac{g_m}{\omega_T} - C_\mu = \frac{0.0385}{2\pi \, (300 \times 10^6)} - 4 \times 10^{-12} = 16.4 \, \text{pF}.$$

The Miller's equivalent input capacitance of the BJT is,

$$\begin{aligned} C_i &= C_\pi + C_\mu \, (1 + g_m R) = C_\pi + C_\mu \, \{1 + g_m \, [R_{B1} \, // \, R_{B2} \, // \, R_C \, // \, (r_b + r_\pi)]\} \\ &= [\, 16.4 + 4 \, [\, 1 + 0.0385 \, [\, 120 \times 10^3 // 120 \times 10^3 // 10^3 // \, (30 + 5.2 \times 10^3)]]] \times 10^{-12} \\ &= 147.8 \, \text{pF}. \end{aligned}$$

There is considerable freedom in the choice of component values Let $C_1' = 0.051 \, \mu\text{F}$. Then,

$$C_1 = C_1' + C_i = 0.051 \, \mu\text{F} + 147.8 \, \text{pF} = 51.15 \, \text{nF}.$$

If the magnitude of the closed loop gain is designed for 1.25, then

$$\frac{\beta_F R_C}{1.25(R_C + r_\pi)} = \frac{C_1}{C_2}$$

$$C_2 = \frac{1.25 \ (R_C + r_\pi)}{\beta_F R_C} = \frac{1.25 \ [\ 1000 + \ 5200] \ (51.15 \times \ 10^{-9})}{200 \ (1000)} = 1.982 \text{ nF}.$$

Select $C_2 = 2000 \text{ pF}$. The frequency of oscillation is 3.5795 MHz as expressed as,

$$f_o = \frac{1}{2\pi} \sqrt{\frac{C_1 + C_2}{LC_1 C_2}}.$$

Solving for L,

$$L = \frac{C_1 + C_2}{(2\pi f_o)^2 C_1 C_2} = \frac{(51.15 \times 10^{-9} + 200 \times 10^{-9})}{[2\pi(3.5795 \times 10^6)]^2 (51.15 \times 10^{-9})(200 \times 10^{-9})}$$
$$= 1.027 \ \mu\text{H} \approx 1 \ \mu\text{H}.$$

Simulation using these values results in an oscillation frequency, $f_o = 3.52 \text{ MHz}$. A simple method for insuring stable and accurate oscillation frequency is to use a 3.5795 MHz crystal. A crystal Colpitts oscillator can be designed by simply replacing the inductor in the reactive feedback network using the calculated capacitor values.

12.6   PROBLEMS

12.1.   For the network shown, show that:

(a) When used with an OpAmp to form an oscillator, the resonant frequency is

$$f = 2\pi RC \quad \text{and that the gain must exceed 3.}$$

(b) Use SPICE to verify the result of a).

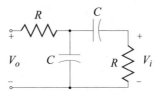

12.2.   Design an OpAmp-based phase shift oscillator circuit at 8 kHz. Assume an ideal OpAmp. Simulate your design to verify proper operation.

12.3.   Show that a four section RC network reduces the required network voltage gain for a phase-shift oscillator.

(a) Determine the minimum required voltage gain of the amplifier.

(b) Determine the resonant frequency of a four section RC phase-shift oscillator.

12.4. Determine the resonant frequency of a phase-shift oscillator with a three RC sections with capacitors to ground as shown.

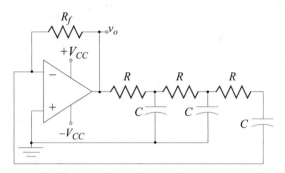

12.5. For a four section RC phase-shift network, similar to Problem 12.4 with capacitors to ground, for a phase-shift oscillator:

(a) Determine the minimum required voltage gain of the amplifier.

(b) Determine the resonant frequency of the phase-shift oscillator.

12.6. Given FET parameters:

$$V_{PO} = -1.5\,\text{V}, \quad I_{DSS} = 12\,\text{mA},$$
$$V_A = 200\,\text{V}.$$

(a) Complete and modify the design shown for an oscillation frequency of 5 kHz. Make any necessary assumptions.

(b) Simulate the circuit using SPICE and confirm the oscillation frequency.

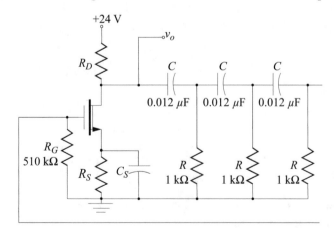

12.7. Given the following BJT parameters of interest: $\beta_F = 200$, and $V_A = 200\,\text{V}$:

(a) Complete the design of the BJT-based phase-shift oscillator shown. What is the frequency of oscillation? Make any necessary assumptions.

(b) Simulate the circuit using SPICE and confirm the oscillation frequency.

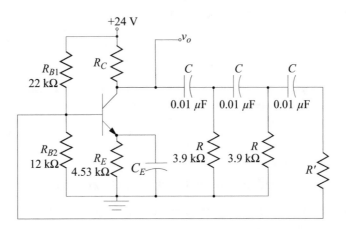

12.8. Complete the design an OpAmp-based Wien-bridge oscillator for 19.2 kHz. Use $C = 1\,\mu\text{F}$ and $R_f = 5.1\,\text{k}\Omega$.

Simulate the circuit using SPICE and confirm the oscillation frequency.

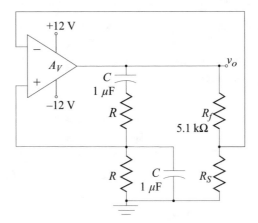

12.9. Using an ideal OpAmp, design a Colpitts oscillator at 1 kHz using an inductor value of $L = 10\,\mu\text{H}$. Simulate the completed design and verify the oscillation frequency.

12.10. Design a BJT-based common-emitter configured Colpitts oscillator at 1 MHz. A +15 V power supply is available. The BJT parameters of interest are:

$$\beta_F = 200, \ V_A = 150\,\text{V}, r_b = 30\,\Omega, \ C_{ibo} = 8\,\text{pF} \quad \text{at} \quad V_{EB} = 0.5\,\text{V},$$
$$C_{obo} = 4\,\text{pF} \quad \text{at} \quad V_{CB} = 5\,\text{V}, \quad \text{and} \quad f_T = 300\,\text{MHz} \quad \text{at} \quad I_C = 10\,\text{mA}.$$

Simulate the circuit using SPICE and confirm the oscillation frequency.

12.11. Given the following BJT parameters:

$$\beta_F = 200, \ V_A = 150\,\text{V}, \ r_b = 30\,\Omega, \ C_{ibo} = 8\,\text{pF} \quad \text{at} \quad V_{EB} = 0.5\,\text{V},$$
$$C_{obo} = 4\,\text{pF} \quad \text{at} \quad V_{CB} = 5\,\text{V}, \quad \text{and} \quad f_T = 300\,\text{MHz} \quad \text{at} \quad I_C = 10\,\text{mA}.$$

(a) Design a BJT-based common-emitter configured Colpitts oscillator at 10 MHz. Bias the transistor at $I_C = 1\,\text{mA}$ and $V_{CE} = 8\,\text{V}$. A power supply voltage of +24 V is available.

(b) Re-design the oscillator using the parameters found in Table 12.1 for a 10 MHz fundamental crystal.

(c) Simulate both circuits with SPICE and comment on the results.

12.12. Design an NJFET-based common-source configured Colpitts oscillator at 2 MHz. The FET parameters of interest are:

$$I_{DSS} = 6\,\text{mA}, \ V_{PO} = -4.7\,\text{V}, \ V_A = 100\,\text{V},$$
$$C_{iss} = 4.5\,\text{pF} \quad \text{at} \quad V_{GS} = 0\,\text{V}, \quad \text{and} \quad C_{rss} = 1.5\,\text{pF} \quad \text{at} \quad V_{GS} = 0\,\text{V}.$$

Assume a +15 V power supply. Bias the transistor at $I_D = 1\,\text{mA}; V_{DS} = 5\,\text{V}$.

Simulate the circuit using SPICE and confirm the oscillation frequency.

12.13. Determine the relationship for the frequency of oscillation for the common-base configured Colpitts oscillator in terms of C_1, C_2, and L using small-signal hybrid-π analysis and appropriate assumptions. Assume that $r_o^{-1} \ll C_1 C_2' (r_b + r_\pi)$, and C_C and C_B are large-valued capacitors.

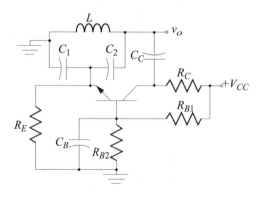

12.14. Determine the relationship for the frequency of oscillation for the common-gate configured Colpitts oscillator in terms of C_1, L using small-signal analysis and appropriate assumptions. C_D and C_G are large-valued capacitors.

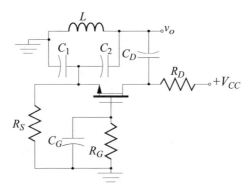

12.15. Design a BJT-based common-base configured Colpitts oscillator at 1 MHz. The BJT parameters of interest are:

$$\beta_F = 200, \ V_A = 150\,\text{V}, \ r_b = 30\,\Omega, \ C_{ibo} = 8\,\text{pF} \quad \text{at} \quad V_{EB} = 0.5\,\text{V},$$
$$C_{obo} = 4\,\text{pF} \quad \text{at} \quad V_{CB} = 5\,\text{V}, \quad \text{and} \quad f_T = 300\,\text{MHz} \quad \text{at} \quad I_C = 10\,\text{mA}.$$

Assume a +15 V power supply. Bias the transistor at $I_C = 1\,\text{mA}; V_{CE} = 5\,\text{V}$.

Simulate the circuit using SPICE and confirm the oscillation frequency.

12.16. Design an NMOSFET-based common-gate configured Colpitts oscillator at 2 MHz. The FET parameters of interest are:

$$I_{DSS} = 6\,\text{mA}, \ V_{PO} = -4.7\,\text{V}, \ V_A = 150\,\text{V},$$
$$C_{iss} = 4.5\,\text{pF} \quad \text{at} \quad V_{GS} = 0\,\text{V}, \quad \text{and} \quad C_{rss} = 1.5\,\text{pF} \quad \text{at} \quad V_{GS} = 0\,\text{V}.$$

Assume a +15 V power supply. Bias the transistor at $I_D = 1\,\text{mA}; V_{DS} = 5\,\text{V}$.

Simulate the circuit using SPICE and confirm the oscillation frequency.

12.17. Design an NJFET-based common-source configured Hartley oscillator at 1 MHz. The FET parameters of interest are:

$$I_{DSS} = 10\,\text{mA}, \ V_{PO} = -3.5\,\text{V}, \ V_A = 150\,\text{V},$$
$$C_{iss} = 6\,\text{pF} \quad \text{at} \quad V_{GS} = 0\,\text{V}, \quad \text{and} \quad C_{rss} = 2\,\text{pF} \quad \text{at} \quad V_{GS} = 0\,\text{V}.$$

Assume a +15 V power supply. Bias the transistor at $I_D = 1\,\text{mA}; V_{DS} = 5\,\text{V}$.

Simulate the circuit using SPICE and confirm the oscillation frequency.

12.18. Determine the relationship for the frequency of oscillation for the transformer coupled common-base configured Hartley oscillator shown in terms of C, L, and the transformer turns ratio using small-signal hybrid-π analysis and appropriate assumptions. Assume that C_C and C_B are large-valued capacitors.

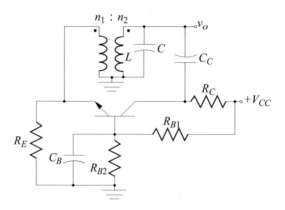

12.19. In many instances it is preferable to use auto-transformers, which are easily fabricated, instead of regular transformers for coupling signals in feedback configurations. The auto-transformer and its equivalent circuit are shown in the figure:

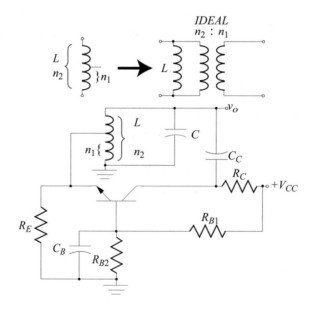

Determine the relationship for the frequency of oscillation for the auto-transformer coupled common-base configured Hartley oscillator shown in terms of C, L, and the auto-transformer turns ratio using small-signal hybrid-π analysis and appropriate assumptions. Assume that C_C and C_B are large-valued capacitors.

12.20. What is the relationship for the frequency of oscillation for the auto-transformer coupled common-base configured Hartley oscillator shown in Problem 12.16 when R_E is eliminated ($R_E \rightarrow \infty$)?

12.21. Design a BJT-based auto-transformer coupled common-base configured Hartley oscillator at 1 MHz. The BJT parameters of interest are:

$$\beta_F = 200, \ V_A = 150\,\text{V}, \ r_b = 30\,\Omega, \ C_{ibo} = 8\,\text{pF} \quad \text{at} \quad V_{EB} = 0.5\,\text{V},$$
$$C_{obo} = 4\,\text{pF} \quad \text{at} \quad V_{CB} = 5\,\text{V}, \quad \text{and} \quad f_T = 300\,\text{MHz} \quad \text{at} \quad I_C = 10\,\text{mA}.$$

Assume a $+15\,\text{V}$ power supply. Bias the transistor at $I_C = 1\,\text{mA}$; $V_{CE} = 5\,\text{V}$. Simulate the circuit using SPICE and confirm the oscillation frequency.

12.22. Determine the relationship for the frequency of oscillation for the transformer coupled common-source configured Hartley oscillator shown in terms of C, L, and the auto-transformer turns ratio using small-signal analysis and appropriate assumptions. Assume that C_C and C_G are large-valued capacitors.

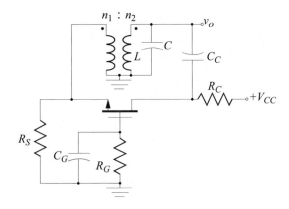

12.23. Determine the relationship for the frequency of oscillation for the auto-transformer coupled common-source configured Hartley oscillator shown in terms of C, L, and the auto-transformer turns ratio using small-signal analysis and appropriate assumptions. Assume that C_C and C_G are large-valued capacitors.

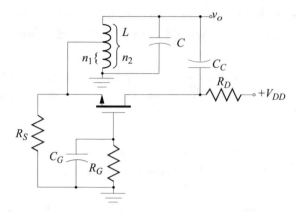

12.24. What is the relationship for the frequency of oscillation for the auto-transformer coupled common-gate configured Hartley oscillator shown in Problem 12.19 when R_S is eliminated ($R_S \to \infty$)?

12.25. Determine the relationship for the frequency of oscillation for the auto-transformer coupled common-collector configured Hartley oscillator shown in terms of $C, L,$ and the auto-transformer turns ratio using small-signal hybrid-π analysis and appropriate assumptions.

Assume that C_C and C_B are large-valued capacitors.

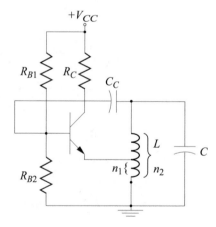

12.26. Design an NMOSFET-based auto-transformer coupled common-gate configured Hartley oscillator at 2 MHz. The FET parameters of interest are:

$$I_{DSS} = 6\,\text{mA}, \; V_T = 2.0\,\text{V}, \; V_A = 150\,\text{V},$$
$$C_{iss} = 4.5\,\text{pF} \quad \text{at} \quad V_{GS} = 0\,\text{V}, \quad \text{and} \quad C_{rss} = 1.5\,\text{pF} \quad \text{at} \quad V_{GS} = 0\,\text{V}.$$

Assume a $+15$ V power supply. Bias the transistor at $I_D = 1$ mA; $V_{DS} = 5$ V.

Simulate the circuit using SPICE and confirm the oscillation frequency.

12.27. If the temperature stability of the crystal is 5.0 ppm/°C, determine the percent change in oscillator frequency after a rise in temperature of 50°C for a Pierce oscillator. Assume that the temperature variation only affect the series equivalent capacitance of the crystal.

12.28. A small inductance is added in parallel with a crystal operating in the series mode in a crystal oscillator. Will the frequency of the oscillation increase or decrease? Explain. Use simulations where appropriate.

12.29. A small "trimmer" capacitance is added in parallel with a crystal operating in the antiresonant mode in a crystal oscillator. Will the frequency of the oscillation increase or decrease? Explain. Use simulations where appropriate.

12.30. Design a test circuit to confirm the resonance frequency of a 2 MHz crystal with the parameters given in Table 12.1. Confirm its operation by simulating the circuit using SPICE.

12.31. Design a 1 MHz CMOS Pierce oscillator using the crystal data in Table 12.1. The MOS parameters are:

$$K_n = 100 \, \mu A/V^2, \ K_n = 100 \, \mu A/V^2,$$
$$V_{Tn} = |V_{Tn}| = 1 \text{ V}, \ C_{iss} = 50 \text{ pF} \quad \text{at} \quad V_{GS} = 0 \text{ V}, \quad \text{and} \quad C_{rss} = 5 \text{ pF} \quad \text{at} \quad V_{GS} = 0 \text{ V}.$$

12.32. Find the condition under which the circuit shown will oscillate.

Assume $V_A = 180$ V.

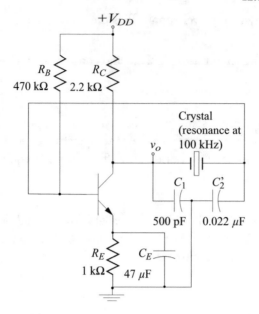

REFERENCES

[1] Ghausi, M. S., *Electronic Devices and Circuits: Discrete and Integrated*, Holt, Rinehart and Winston, New York, 1985.

[2] Millman, J. and Halkias, C. C., *Integrated Electronics: Analog and Digital Circuits and Systems*, McGraw-Hill Book Company, New York, 1972.

[3] Sedra, A. S. and Smith, K. C., *Microelectronic Circuits*, 3rd ed., Holt, Rinehart, and Winston, Philadelphia, 1991.

[4] Schilling, D. L. and Belove, C., *Electronic Circuits*, 3rd ed., McGraw-Hill Book Company, New York, 1989.

[5] Young, P. H., *Electronic Communication Techniques*, 3rd ed., Merrill Publishing Company, New York, 1994.

CHAPTER 13

Waveform Generation and Waveshaping

In addition to sinusoidal waveforms, electronic systems often have need for signals with other wave shapes. Common waveforms include single pulses of fixed duration as well as periodic square waves and triangular waves. Fixed-duration pulses are used primarily for timing of events in communication and control systems. Periodic square waves are formed by a regular series of pulses: one typical use is as a clock for digital systems. If pulse timing, duration, or amplitude can be externally controlled, these altered square waves become the foundation of many digital communication systems. Triangular waves are particularly significant in scanning an electron beam across a cathode-ray tube screen (as in television or oscilloscope applications), in precise time measurements, and in time modulation.

Electronic circuits that generate non-sinusoidal waveforms, such as pulse, square, and triangular waveforms, are typically based upon electronic multivibrators. These multivibrators are characterized by a very rapid transition between two distinct output states and can be grouped into three basic categories based on the time stability of these output states. A bistable multivibrator will rest indefinitely in either output state until triggered to change state. A monostable multivibrator has one stable state: the other state is of fixed, finite duration (a quasi-stable state) that can only be activated with a triggering signal. A constant, periodic switching between quasi-stable states characterizes an astable multivibrator. Astable, periodic switching is also known as non-linear oscillation.

In this chapter, multivibrator circuits intended for precision analog applications are presented. These circuit designs are based on a slight variation of the operational amplifier, typically known as the *comparator*. Circuits that generate single pulses of varying time duration or square and triangular waveforms of variable time symmetry using multivibrators are described. Similar circuits utilizing an integrated-circuit multivibrator, in the form of an IC timer, are also described. A voltage-controlled square-wave oscillator (VCO) is presented as a special case of non-linear oscillation.

Arbitrary, periodic waveforms are often derived from triangular waveforms. Two techniques to generate sinusoids from triangular waveforms are described as examples of waveform alteration. The primary technique is based on piece-wise linear amplification using the forms of the basic diode clipping circuits described in Chapter 2 (Book 1).

Multivibrator applications based on OpAmps and comparators are best suited for electronic applications from a fraction of a Hertz up to a few Megahertz. High-frequency, analog multivibrator applications are beyond the scope of this discussion. High-frequency multivibrators intended for, digital applications are presented in Chapter 16.

13.1   MULTIVIBRATORS

Multivibrator circuits are fundamental to many waveshaping and wave generation circuits. In the most common form they have two output states, each of which may be either stable or quasi-stable. As such, multivibrator circuits can be grouped into three classifications based on the stability of the output states:

- *Bistable Multivibrator*—two stable output states. Bistable circuits require a triggering signal to transition between output states. Once in a particular output state, the circuit remains in that stable state indefinitely until triggered for a transition to the other stable output state.

- *Astable Multivibrator*—two quasi-stable output states. Without any external triggering, the astable multivibrator transitions periodically between two quasi-stable output states.

- *Monostable Multivibrator*—one stable and one quasi-stable output state. In monostable circuits, a triggering signal is required to induce an output transition from the stable state to the quasi-stable state. After the circuit remains in the quasi-stable state for a fixed time, typically long in comparison with the time of transition between states, it returns to the stable state without external triggering.

13.1.1   BISTABLE MULTIVIBRATORS

One of the most simple forms of a bistable circuit is the comparator. A comparator shares many characteristics with a high voltage gain OpAmp: it has two inputs and a single output determined by the difference of two input signals. As seen in Figure 13.1, the circuit symbol for a comparator is identical with that of an OpAmp and the terminals are identified in the same manner. If the non-inverting input is at a higher potential than the inverting input, the output of the comparator is the HIGH voltage output state: reversing the sense of the inputs yields the LOW voltage output state. The HIGH and LOW output voltage states are often determined by the power supply rails, but may be internally controlled to be other voltage levels (5 V and 0 V are a common pair when interfacing to digital circuitry). While OpAmps can be used as comparators, a comparator is not typically used with a wide range of negative feedback. Therefore, the comparator can be optimized for rapid transition between states at the expense of linearity and feedback stability. As such, a comparator will typically transition between states much more quickly than an OpAmp of similar design. Comparators suffer from the same non-ideal characteristics as OpAmps. Slew rate and frequency response are, arguably, the most problematic of these non-ideal properties.[1]

---

[1]Non-ideal characteristics of OpAmps are discussed in Section 1.5 (Book 1). The discussion is also valid for comparators.

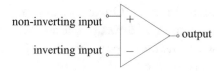

Figure 13.1: Comparator circuit symbol.

While comparators perform well in a noiseless environment, the noise content of a typical signal can create false triggering or multiple triggering of the output of a comparator. In a typical application, a comparator is used to determine when the non-inverting input voltage exceeds the voltage level at the inverting input. As shown in Figure 13.2, the addition of noise to an increasing input signal at the non-inverting input can cause several transitions of the comparator output when the input voltage nears the voltage reference level present at the inverting input. Usually a single transition of the output is desired in such a comparison.

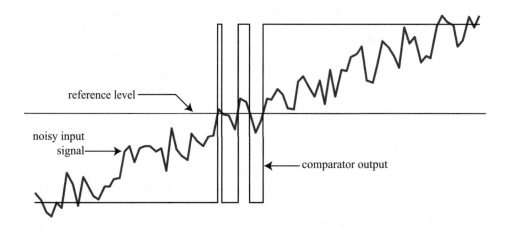

Figure 13.2: The output of a comparator due to a noisy input.

Multiple triggering of the output of a comparator can be eliminated with the application of *positive* feedback. Positive feedback retains the rapid output transition of comparator, but alters the trigger level so that two separate trigger levels exist: one for positive slope signals and another for negative slope signals. The resultant transfer relationship exhibits hysteresis as shown in Figure 13.3. Any input signal below the negative slope transition voltage, V_T^-, results in a LOW output, V_L. If the output state is LOW, it will not transition to the HIGH state unless the input is greater than the positive slope transition voltage, V_T^+. Similarly, any input above V_T^+ results in a HIGH output, V_H, that will not transition to the LOW state unless the input falls below

V_T^-. Thus, signals, after crossing a threshold, do not respond to input signal changes unless the variation is large enough to cross the deadband.

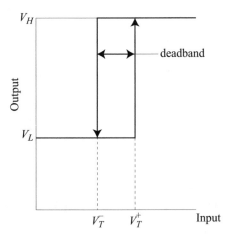

Figure 13.3: Schmitt trigger transfer characteristic hysteresis.

An example of a comparator with positive feedback is shown in Figure 13.4. This form of circuit is called a *Schmitt trigger*. Aside from being especially useful in converting slowly varying or a noisy signal into a clean, pulsed form with sharp transitions, the Schmitt trigger is particularly useful in converting sine-wave input into a pulse-train output. Variation of the pulse-train duty cycle in this application is accomplished by varying the triggering voltage levels, V_T^+ and V_T^-.

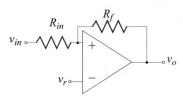

Figure 13.4: A simple Schmitt trigger circuit.

The Schmitt trigger can be realized in many configurations. When comparators are used as the basic active element,[2] inverting and non-inverting forms are typically realized as is shown in Figure 13.5. The additional resistors shown provide stable reference voltage, v_r. If a more precise reference voltage is necessary, precision voltage reference circuitry may be utilized instead of simple resistor networks.[3]

---

[2]Bipolar and CMOS realizations of Schmitt trigger circuits for digital applications are presented in Section 16.5.
[3]Precision voltage references are discussed in Section 14.2.1.

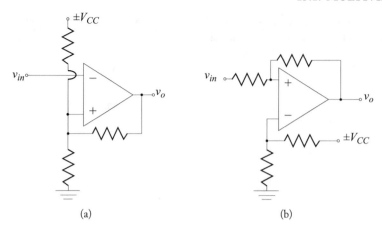

Figure 13.5: Typical Schmitt trigger circuits. (a) Inverting Schmitt trigger; (b) Non-inverting Schmitt trigger.

Example 13.1   Design Example

The simple Schmitt trigger circuit of Figure 13.4 is configured so that the two stable output voltages are ± 10 V. The reference voltage, v_r, is a ground potential. Complete the design by choosing R_{in} and R_f so that the output transition to $+10$ V occurs at the threshold voltage, $v_{in} = V_T^+ = 0.5$ V. At what input voltage level does the transition to -10 V (that is, V_T^-) occur?

Solution:

The output will transition to $+10$ V when the non-inverting input reaches the reference voltage, $v_r = 0$ V. The non-inverting input voltage can be obtained from the input voltage, $v_{in} = 0.5$ V and the LOW output voltage, $v_o = -10$ V, using the voltage division technique:

$$v_r = -10 + \frac{R_f}{R_f + R_{in}}[0.5 - (-10)] = 0.0,$$

or

$$\frac{R_f}{R_f + R_{in}} = \frac{10}{10.5} \quad \Rightarrow \quad R_f = 20\, R_{in}.$$

Many resistor pairs will fulfill this simple requirement. As with OpAmps, it is important to choose resistors that are small compared to the input resistance of the comparator and large compared to the output resistance. The following choice is one pair of values that will meet the design goals using standard resistance values:

$$R_f = 20\,\text{k}\Omega \quad \text{and} \quad R_{in} = 1\,\text{k}\Omega.$$

The transition to -10 V is an exact mirror image of the transition to $+10$ V: it occurs at

$$v_{in} = -0.5\,\text{V}.$$

Thus, the HIGH output state $(+10\,\text{V})$ is triggered when the input exceeds 0.5 V and remains until the input falls below -0.5 V.

13.1.2 ASTABLE MULTIVIBRATORS

Astable multivibrators continuously transition between quasi-stable output states without the aid of an external triggering input. The output of an astable therefore becomes a square wave with the waveform duty cycle as a possible variable. One such astable circuit is shown in Figure 13.6. Here the two inputs to a comparator are coupled to the comparator output through two different type networks. The non-inverting input is connected through a resistive voltage divider: as the output toggles between quasi-static states, this input similarly toggles between states. The inverting input is connected to the comparator output by an RC charging network: this input exponentially transitions toward the output voltage. When the inverting input voltage matches the voltage at the non-inverting input, the output voltage toggles to the other quasi-stable state. This output toggling causes successive exponential change and toggling.

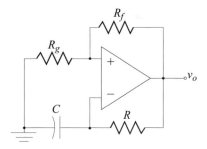

Figure 13.6: An astable multivibrator.

The two quasi-stable output voltage states of the astable multivibrator are at fixed voltages, V_H and V_L. When the output is in one of these states, the voltage state at the non-inverting ("+") comparator node is given by:

$$v_H^+ = \frac{R_g}{R_g + R_f} V_H \quad \text{or} \quad v_L^+ = \frac{R_g}{R_g + R_f} V_L. \tag{13.1}$$

Just prior to a HIGH to LOW transition, for example, the output voltage is V_H, the non-inverting input is at v_H^+, and the inverting input is in exponential transition from its initial value, v_L^+, toward V_H. This exponential transition will continue until the voltage at the non-inverting input matches that at the inverting input, v_H^+, at which time the output will toggle to V_L. That is,

$$v^-(t) = V_H - \left(V_H - v_L^+\right) e^{-\frac{t}{RC}}. \tag{13.2}$$

The total exponential transition time is the solution to the expression

$$v^- (t) = v_H^+,\tag{13.3}$$

or

$$V_H - \left(V_H - \frac{R_g}{R_g + R_f}V_L\right) e^{-\frac{t}{RC}} = \frac{R_g}{R_g + R_f}V_H.\tag{13.4}$$

The transition time is given by:

$$t_{HL} = RC \ln \left\{\frac{(R_g + R_f) V_H - R_g V_L}{R_f V_H}\right\} = RC \ln \left\{1 + \frac{R_g (V_H - V_L)}{R_f V_H}\right\}.\tag{13.5}$$

A LOW to HIGH transition time is determined in the same manner: the expression is the same as Equation (13.5) with V_H and V_L interchanged. As such, the duty cycle of the output square wave can be altered somewhat. The period of this repeated output toggling becomes the sum of the two transition times:

$$\tau = RC \left[\ln \left\{1 + \frac{R_g (V_H - V_L)}{R_f V_H}\right\} + \ln \left\{1 + \frac{R_g (V_L - V_H)}{R_f V_L}\right\}\right].\tag{13.6}$$

The complexity of these expressions for transition time can be simplified greatly by assuming a particular comparator operational configuration. Typically, an astable multivibrator is operated with symmetric voltage limits, $V_L = -V_H$. This operation practice results in identical transition times and a 50% duty cycle:

$$t_{HL} = t_{LH} = RC \ln \left\{1 + 2\frac{R_g}{R_f}\right\}.\tag{13.7}$$

In this case, the period of the repeated output toggling is twice the transition time:

$$\tau = 2 t_{HL} = 2RC \ln \left\{1 + 2\frac{R_g}{R_f}\right\}.\tag{13.8}$$

In practice, the slew rate limitations of the circuit comparator will lengthen each of the transition times lowering the frequency of operation somewhat. The waveforms associated with a symmetric astable multivibrator are shown in Figure 13.7.

13.1.3 MONOSTABLE MULTIVIBRATORS

A monostable multivibrator produces a single output pulse, typically of precise amplitude and duration, each time a trigger signal is applied to the input. As such, monostable multivibrators are useful in transforming a train of pulses with variable amplitude and/or duration into a train of pulses with standard amplitude and duration. The output during the single pulse is quasi-stable: between pulses the output is stable.

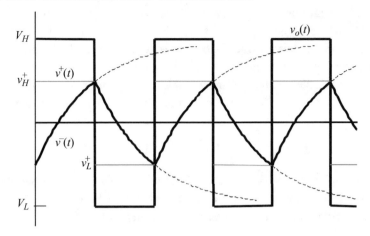

Figure 13.7: Astable multivibrator waveforms.

An astable multivibrator can be transformed into a monostable multivibrator by stabilizing one of the quasi-stable output states. The circuit of Figure 13.8 has its negative output, V_L, stabilized by the introduction of the stabilizing diode, D_S. This diode prevents the inverting input, $v^-(t)$, from becoming sufficiently negative to fall below the lower non-inverting state:

$$v^-(t) > v_L^+ = \frac{R_g}{R_g + R_f} V_L. \tag{13.9}$$

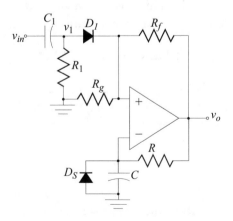

Figure 13.8: A monostable multivibrator.

Therefore, the comparator will not spontaneously toggle to the positive output state, V_H. In order for an external signal to trigger a pulse, the capacitive input circuit formed by C_1, R_1, and D_1 is added. This input circuit allows a fast rise time pulse to momentarily lift the non-inverting input above the inverting input which triggers a HIGH output.

As with the astable circuit, the monostable multivibrator output states are given by V_H (quasi-stable) and V_L (stable). When the circuit is in the stable state, the two inputs to the comparator are:

$$v_L^+ = \frac{R_g}{R_g + R_f}\, V_L \quad \text{and} \quad v_L^- = -V_\gamma \approx -0.7 \text{ V.} \tag{13.10}$$

A positive pulse applied to the input, v_{in}, will raise the non-inverting input above the inverting input and force the comparator output to the HIGH state, V_H. The non-inverting input toggles and the inverting input begins an exponential transition to V_H:

$$v_H^+ = \frac{R_g}{R_g + R_f}\, V_H,$$

and $\hspace{11cm}$ (13.11)

$$v^-(t) = V_H - \left(V_H - v_L^-\right) e^{-\frac{t}{RC}}.$$

The exponential transition will continue until the voltage at the non-inverting input matches that at the inverting input, v_H^+, at which time the output will toggle to V_L. The duration of the positive pulse is given by the time of this exponential transition:

$$t_{pulse} = RC \ln \left\{ \frac{\left(R_g + R_f\right)\left(V_H - v_L^-\right)}{R_f V_H} \right\}. \tag{13.12}$$

After the output toggles LOW, the non-inverting input returns to its stable state, v_L^+, and the inverting input begins an exponential transition toward V_L:

$$v^-(t) = V_L - \left(V_L - v_H^+\right) e^{-\frac{t}{RC}}. \tag{13.13}$$

This exponential transition is halted by the diode, D_S, before the inverting input is sufficiently negative to toggle the comparator HIGH. Consequently, the LOW output state is stable. The time to return to the stable state is given by:

$$t_{recovery} = RC \ln \left\{ \frac{V_L - v_H^+}{V_L - v_L^-} \right\}. \tag{13.14}$$

Once the monostable multivibrator returns to its stable state, it can be retriggered to output another single pulse. The characteristic of a single output pulse for each triggering leads to the common alternate identification of a monostable multivibrator as a *one-shot*. Typical monostable

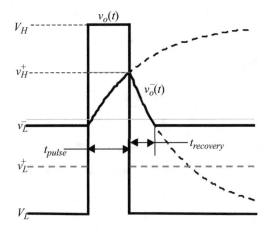

Figure 13.9: Monostable multivibrator waveforms.

multivibrator waveforms are shown in Figure 13.9. In order to ensure clean, single pulses, design guidelines suggest that the input circuit time constant be small compared to the pulse duration. One-shot circuits find greatest use in analog or asynchronous digital circuitry. Use in synchronous digital circuitry is discouraged due to a variety of problems.

Example 13.2   Design Example

The monostable multivibrator circuit of Figure 13.8 is configured so that the two stable output voltages are $\pm 10\,\text{V}$. Complete the design so that the circuit responds to a pulse input with an output pulse of duration $10\,\text{ms}$. What is the recovery time for the design?

Solution:

The pulse duration is given by Equation (13.12):

$$t_{pulse} = RC \ln \left\{ \frac{(R_g + R_f)(V_H - v_L^-)}{R_f V_H} \right\} = RC \ln \left\{ \frac{(R_g + R_f)(10.7)}{R_f 10} \right\}.$$

In order to reduce the effects of stray noise, it is good practice to keep within the first two time constants of the exponential decay. While many sets of component values will satisfy the constraints, the following set is chosen.

Choose $RC = t_{pulse}$ (one time constant) $= 10\,\text{ms}$.

One pair of standard components meeting this choice are:

$$R = 10\,\text{k}\Omega \quad \text{and} \quad C = 1\,\mu\text{F},$$

a further consequence of this choice is:

$$1 = \ln \left\{ \frac{(R_g + R_f)\,(10.7)}{R_f\,10} \right\}.$$

One pair of standard value resistors that will meet this requirement are:

$$R_g = 15.4\,\text{k}\Omega \quad \text{and} \quad R_f = 10\,\text{k}\Omega.$$

The recovery time is determined by Equation (13.14). With the above choices for component values it is:

$$t_{recovery} = 10\,\text{k}\,(1\,\mu) \ln \left\{ \frac{-10 - 5}{-10 - (-0.7)} \right\} = 4.78\,\text{ms}.$$

13.2 GENERATION OF SQUARE AND TRIANGULAR WAVEFORMS USING ASTABLE MULTIVIBRATORS

In electronic applications, the three most useful waveforms are:

- sinusoidal waveforms,

- square waveforms, and

- triangular waveforms.

Precision generation of these waveforms is vital to the proper operation of a wide variety of electronic devices and electronic test equipment. Square wave can be generated using by either passing a sinusoid through a bistable circuit, such as a Schmitt trigger, or directly with an astable multivibrator. Triangular waveforms can be obtained by integrating a square waveform. Simultaneous generation of all three waveform types presents a challenge. Fortunately one form of astable multivibrator is capable of simultaneous generation of square and triangular waveforms: sinusoids can be derived from the triangular waveform using non-linear waveshaping techniques.[4]

13.2.1 PRECISION SQUARE WAVE GENERATION

The astable multivibrator discussed in Section 13.1.2 is a near-ideal generator of square waves. With symmetric output voltage levels, a uniform square wave is generated with period,

$$\tau = 2RC \ln \left\{ 1 + 2\frac{R_g}{R_f} \right\}. \tag{13.15}$$

[4]These non-linear techniques are discussed in Section 13.3.

A precision square-wave generator based on this astable multivibrator is shown in Figure 13.10. In this circuit the symmetric output voltage levels are ensured through the use of a back-to-back, matched pair of Zener diodes.

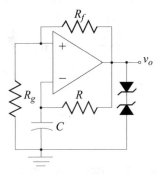

Figure 13.10: An astable multivibrator square wave generator.

Alteration of the time symmetry of the waveform so that the HIGH output time is not equal to the LOW output time is often desired in such a circuit. Two possible techniques used to alter the symmetry are significant in any discussion of multivibrators:

- A monostable multivibrator (one-shot) is connected in series with the output of the circuit of Figure 13.10. The monostable multivibrator is adjusted to output a pulse of varying duration without varying the period of the waveform.

- The astable multivibrator circuit is modified so that the HIGH and LOW output times are unequal.

Due to the significant increase in the complexity of the circuit, including the addition of another comparator, the addition of a one-shot is usually not the best choice: minor modification of the astable multivibrator is simple and effective.

While it has been shown that non-symmetric HIGH and LOW voltage states will lead to non-symmetric waveforms, it is more common to vary the square-wave time symmetry through control of the exponential RC time constant. The circuit shown in Figure 13.11 is one possible multivibrator realization that provides different charging and discharging time constants. During the HIGH output state, diode, D, conducts and diode, D', is OFF. The circuit reduces to that of Figure 13.10: the time constant is given by the product of RC. The addition of a diode in the negative-feedback path reduces the voltage apparent to the charging network by V_γ. Since the charging time is not dependent on this voltage, the HIGHstate duration time is, as previously derived:

$$t_H = RC \ln \left\{ 1 + 2\frac{R_g}{R_f} \right\}. \tag{13.16}$$

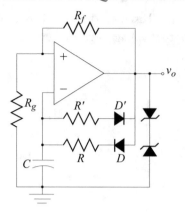

Figure 13.11: An astable, non-symmetric square-wave generator.

When the output is in the LOW state, the operational modes of the two diodes, D and D', are interchanged and the alternate negative-feedback resistor, R', acts in the exponential decay. The resultant LOW output state duration time is similarly given by:

$$t_L = R'C \ln \left\{ 1 + 2\frac{R_g}{R_f} \right\}. \tag{13.17}$$

If the two negative-feedback resistors are not equal in value, $R \neq R'$, the HIGH and LOW duration times are unequal and the square wave is asymmetric. The period of the asymmetric square wave is given by the sum of the two duration times:

$$\tau = t_H + t_L = \left(R + R' \right) C \ln \left\{ 1 + 2\frac{R_g}{R_f} \right\}. \tag{13.18}$$

Duty cycle is typically defined as the ratio of the HIGH state duration, t_H, to the period, τ:

$$\text{duty cycle} = \frac{t_H}{\tau} = \frac{R}{R + R'}. \tag{13.19}$$

Example 13.3   Design Example
Design a circuit using an astable multivibrator to produce a 10 kHz square-wave with a 35% HIGH state duty cycle.

Solution:
    When basing timing on RC decay, it is best to stay approximately within the first two time constants. Choosing $R_f = R_g$ will satisfy this guideline. It is also important to choose resistors that will meet standard comparator resistance guidelines. Therefore, within these guidelines,

arbitrarily choose:

$$R_g = R_f = 10\,\text{k}\Omega.$$

Then

$$\tau = \frac{1}{f} = 100\ \mu s = \left(R + R'\right) C \quad (1.0986)$$

$$(R + R')C = 91.024\,\mu s.$$

Arbitrarily choose a convenient value of C that will keep the resistor values, R and R', within comparator resistance constraints. Here choose

$$C = 0.01\,\mu F \quad \Rightarrow \quad R + R' = 9.1024\,\text{k}\Omega.$$

A 35% duty cycle implies

$$\frac{R}{R + R'} = 0.35 \quad \Rightarrow \quad R = 0.53846\ R'.$$

These two constraints result in the final two standard-value resistors:

$$R' = 5.916\,\text{k}\Omega \approx 5.90\,\text{k}\Omega \quad \text{and} \quad R = 3.1858\,\text{k}\Omega \approx 3.20\,\text{k}\Omega.$$

13.2.2 SIMULTANEOUS SQUARE AND TRIANGLE WAVE GENERATION

Astable multivibrators, as described previously (and shown in Figure 13.10), provide a square wave output and a pseudo-triangular waveform as an internal signal. Unfortunately, this internal waveform has voltage transitions that are exponential rather than linear. Transition linearity can be achieved by replacing RC capacitive charging. Two distinct possibilities for linearization of the signal are dominant in typical realizations:

- Current-source fed capacitive charging, or

- Integration of the square wave.

In the first possibility, the negative-feedback resistor, R, is replaced by a constant current device. This current source typically has as its primary element, a FET operating in the saturation region. The second possibility uses an OpAmp integrator to create a linearly varying signal. Since OpAmp integrators invert the signal, the output of the integrator is fed back to the non-inverting terminal of the comparator, as shown in Figure 13.12. In this form, the comparator takes the form of a simple Schmitt trigger.

This astable circuit also has two quasistatic states. They exist for the square wave output, v_{sq}, at fixed voltages, V_H and $-V_H$ (once again, the symmetry is ensured by back-to-back, matched

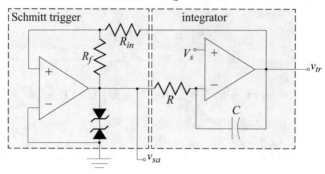

Figure 13.12: Triangular/square wave generator.

Zener diodes). The Schmitt trigger toggles between output states when the voltage at its non-inverting input is at ground potential. This toggle requirement reflects back on the integrator output, v_{tr}, as:

$$v_{tr(toggle)} = -\frac{R_{in}}{R_f} v_{sq} = \mp \frac{R_{in}}{R_f} V_H. \tag{13.20}$$

As previously stated, the time-dependent output, $v_{tr}(t)$, is an inverted integral of the difference between $v_{sq}(t)$ and the symmetry-control voltage, V_s (a constant):

$$v_{tr}(t) = \frac{-1}{RC} \int \left\{ v_{sq}(t) - V_s \right\} dt. \tag{13.21}$$

When v_{sq} is in the HIGH state v_{tr} is linearly decreasing between its toggle values:

$$v_{tr}(t) = \frac{-1}{RC} \int \left\{ V_H - V_s \right\} dt = \frac{V_s - V_H}{RC} t + \text{(a HIGH constant)}. \tag{13.22}$$

The "HIGH constant" of integration need not be evaluated for this discussion. The time for the decreasing voltage transition is given by the difference in toggle values divided by the slope of the linear transition:

$$2 \left(\frac{R_{in}}{R_f} V_H \right) = \frac{V_H - V_s}{RC} t_- \quad \Rightarrow \quad t_- = \frac{2 R_{in} RC V_H}{R_f \left\{ V_H - V_s \right\}}. \tag{13.23}$$

When v_{sq} is in the LOW state v_{tr} is linearly increasing between its toggle values with a different slope

$$v_{tr}(t) = \frac{-1}{RC} \int -\left\{ V_H + V_s \right\} dt = \frac{V_H + V_s}{RC} t + \text{(a LOW constant)}. \tag{13.24}$$

While the difference in toggle values remains the same, the change in slope results in a different transition time for the positive transition:

$$2\left(\frac{R_{in}}{R_f}V_H\right) = \frac{V_H + V_s}{RC}t_+ \quad \Rightarrow \quad t_+ = \frac{2R_{in}RCV_H}{R_f\{V_H + V_s\}}. \tag{13.25}$$

The signal will continuously repeat the transitions. The period for the total waveform is the sum of the transition times:

$$\tau = t_- + t_+ = \frac{4R_{in}RC}{R_f}\left\{\frac{V_H^2}{V_H^2 - V_s^2}\right\}. \tag{13.26}$$

The positive-slope duty cycle is given by the ratio of the positive transition time to the period:

$$\text{duty cycle} = \frac{t_+}{\tau} = \frac{1}{2}\left[1 - \frac{V_s}{V_H}\right]. \tag{13.27}$$

Symmetric waveforms (those with a 50% duty cycle) are obtained when V_s is at ground potential. Positive V_s results the positive-slope waveform segment having shorter duration than the negative-slope segment (positive-slope duty cycle < 50%): negative V_s reverses the relationship. In this circuit, the frequency of oscillation is also dependent on V_s: it is maximized when $V_s = 0$ and decreases in a non-linear fashion as the magnitude of V_s increases. The generator waveforms, v_{tr} and v_{sq}, are shown in Figure 13.13 for positive V_s.

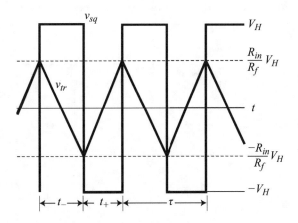

Figure 13.13: Triangle/square wave generator signals.

13.2.3 VOLTAGE-CONTROLLED FREQUENCY OF OSCILLATION

Control of the frequency of oscillation of a waveform generator is often important. The frequency of oscillation for the waveform generator of Section 13.2.2 can be controlled by:

- variation of component values (typically R or C), or

- variation of the symmetry signal, V_s.

A manually-controlled variable resistor or capacitor can be an adequate frequency control device in many applications. Other design constraints may require that the frequency be controlled by electrical, rather than mechanical, inputs. In particular, the need for a voltage-controlled oscillator (VCO) is significant in many demodulation circuits for communication purposes.

While the circuit of Section 13.2.2 meets the voltage-control requirement, there are two significant drawbacks to using this circuit as a VCO:

- a change in the frequency produces a change in the duty cycle

- the frequency depends on the control voltage in a non-linear fashion: the frequency is dependent on $1 - (V_S^2 - V_H^2)$.

Square and triangle waves, with voltage controlled frequency, can be generated simultaneously with a circuit of the basic topology shown in Figure 13.14. This circuit is an astable multivibrator where the rate of linearly charging a capacitor is decoupled from the amplitude of the output of the square wave. This decoupling is accomplished by the insertion of a FET single-pole, double-throw (SPDT) switch between the Schmitt trigger and the integrator. The SPDT switch is controlled by the output of the Schmitt trigger: it alternately couples the input of the integrator to either the frequency-control voltage, v_m, or its negative, $-v_m$ (usually obtained from an OpAmp unity-gain amplifier). It is required that v_m be positive and of sufficient magnitude so that the switch operates properly.

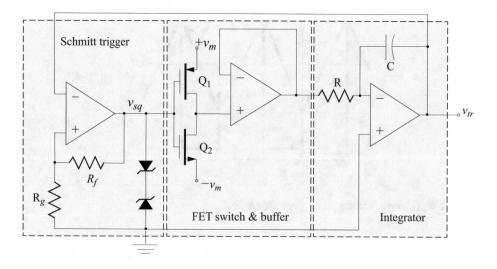

Figure 13.14: A voltage-controlled oscillator.

The output of the integrator, v_{tr}, depends on v_m and v_{sq}:

$$v_{tr}(t) = \frac{-1}{RC} \int \pm v_m \, dt = \frac{\mp v_m}{RC} t + \text{ a constant of integration.} \qquad (13.28)$$

When v_{sq} is HIGH, the output of the FET switch is connected to $-v_m$: when v_{sq} is LOW the output of the switch is connected to $+v_m$. As in the previously discussed waveform generator, the triangle wave transitions are linear with time. The duration of each transition is determined from the toggle voltages and the slope of the transition. The toggle values for the Schmitt trigger are obtained from a voltage division of its two output states: V_H and $-V_H$:

$$v_{tr(toggle)} = \frac{R_g}{R_f + R_g} v_{sq} = \pm \frac{R_g}{R_f + R_g} V_H. \qquad (13.29)$$

The time for the voltage transition is given by the difference in toggle values divided by the slope of the linear transition:

$$\frac{2 R_g}{R_f + R_g} V_H = \frac{v_m}{RC} t \quad \Rightarrow \quad t = \frac{2 R_g RC}{R_f + R_g} \frac{V_H}{v_m}. \qquad (13.30)$$

Circuit symmetry implies that positive and negative transitions are of equal duration: the period of oscillation is double that of a single transition:

$$\tau = 2t = \frac{4 R_g RC}{R_f + R_g} \frac{V_H}{v_m}. \qquad (13.31)$$

The frequency of oscillation is the inverse of the period:

$$f = \frac{1}{\tau} = \frac{R_f + R_g}{4 R_g RC} \frac{v_m}{V_H}. \qquad (13.32)$$

The results of Equation (13.32) indicate the desired linear dependence of frequency with input voltage, v_m. Typically, the linearity extends over three or more decades. If greater frequency variation is necessary, it must be accomplished through resistor switching (typically resistor, R, is switched).

Example 13.4   Design Example
Design a voltage controlled oscillator that will output a 15 V peak-to-peak square wave of variable frequency from 100 Hz to 2 kHz for an input voltage range of 0.1 V to 2 V.

Solution:
The output of the circuit shown in Figure 13.14 will satisfy all design requirements with proper choice of components and component values. A square wave can be obtained at the v_{sq} terminal of the circuit. The 15 V_{p-p} output voltage requirement is obtained by careful choice of

the Zener diode matched pair. The pair is chosen so that $2(V_z + V_\gamma) = 15\,\text{V}$. This requirement will result in diodes with a Zener voltage:

$$V_z \approx 6.8\,\text{V}.$$

The frequency variation is obtained from Equation (13.32). Using the lower range of frequency and voltage, the resultant component constraint is:

$$100 = \frac{R_f + R_g}{4\,R_g\,RC} \frac{0.1}{7.5} \quad \Rightarrow \quad \frac{R_f + R_g}{R_g\,RC} = 30,000.$$

Many sets of component values will satisfy this constraint and keep within the standard comparator resistance guidelines. One set is:

$$R_f = R_g = 10\,\text{k}\Omega \qquad \Rightarrow \qquad RC = 66.67 \times 10 - 6$$
$$C = 4700\,\text{pF} \qquad \Rightarrow \qquad R = 14.2\,\text{k}\Omega.$$

All these values are standard component values.

13.3 NON-LINEAR WAVEFORM SHAPING

Arbitrarily shaped, periodic signals can be derived from triangular waveforms through a two-port network with an appropriate, non-linear transfer function. The non-linear property of the network is usually derived using a piece-wise linear approximation through the introduction of a series of breakpoints. Typically, the breakpoints are realized with diode[5] or transistor switching networks. While selectively filtering the harmonic content of the output of a multivibrator may seem to be an effective, alternate means of waveshaping, this technique is usually not a practical solution. The steep transition regions required to filter the various harmonics imply a very complex, high-order filter design. In most situations the design goals can be more easily met by piece-wise linear approximation. Digitally generated waveforms are another alternative that is particularly popular in arbitrary waveform generators: the design of digital waveform generators is beyond the scope of this discussion.

A general waveshaping network using diodes is shown in Figure 13.15. A triangle wave, v_{tr}, enters the network, is shaped by the piece-wise linear two-port network, and then buffered by the unity gain amplifier: the shaped output is v_o. The network is composed of an array of elements consisting of a series-connected diode, resistor, and voltage sources. If the voltages $\{V_i^+\}$ consist of a progressively increasing set of positive voltages, this portion of the array will progressively flatten the waveform as it increases. Similarly, an increasing magnitude set of negative voltages $\{V_i^-\}$ will progressively flatten the negative portion of the waveform. Reversing a diode in either the upper or lower portion of the array will have the opposite effect on the waveform at the breakpoint.

---

[5]Diode switching networks to shape waveforms are initially introduced in Section 2.6 (Book 1).

The primary difficulty encountered in this form of array is provision for many independent voltage sources. Surprisingly, design solutions are simplified by the switching nature of the array. One possibility is the use of forward-biased diodes to specify the break-point voltages. In order to demonstrate the utility of such an array, a simple triangle-to-sinusoid converter will be demonstrated.

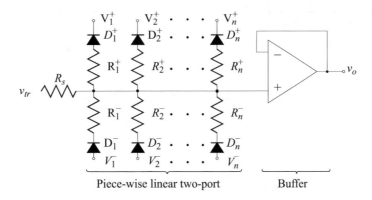

Figure 13.15: A piece-wise linear waveshaping circuit.

Example 13.5   Design Example: A Triangle-to-Sinusoid Converter
In order to produce a sinusoid of magnitude, V_s, from a triangle wave of magnitude, $V_{tr(\max)}$; the transfer function for the triangle-to-sinusoid converter must be given by the expression:

$$v_o = V_s \sin\left(\frac{\pi}{2}\frac{v_{tr}}{V_{tr(\max)}}\right).$$

It has been decided to approximate this transfer function with the seven-segment diode array shown. Here the voltage sources are provided by forward-biased diodes. This array clips all inputs so that the maximum possible output is $\pm 3\,V_\gamma \approx \pm 1.8\,\text{V}$. There are four other breakpoints (progressively introduced by diode conduction) at ±0.6 V and ±1.2 V. The slopes of the transfer function in the various regions are given by:[6]

$$m_1 = 1 \qquad\qquad -0.6\,\text{V} < v_o < 0.6\,\text{V}$$

$$m_2 = \frac{R_1}{R_1 + R_s} \qquad\qquad 0.6\,\text{V} < |v_o| < 1.2\,\text{V}$$

$$m_3 = \frac{R_1 R_2}{R_1 R_2 + R_1 R_s + R_2 R_s} \qquad\qquad 1.2\,\text{V} < |v_o| < 1.8\,\text{V}.$$

The analytic expression for the slope of the transfer function is given by:

$$\frac{d v_o}{d v_{tr}} = \frac{\pi}{2}\frac{V_s}{V_{tr(\text{max})}}\cos\left(\frac{\pi}{2}\frac{v_{tr}}{V_{tr(\text{max})}}\right),$$

slope at the various breakpoints yields:

$$m_1|_{v_{tr}=0} = 1 \quad\Rightarrow\quad \frac{V_s}{V_{tr(\text{max})}} = \frac{2}{\pi}.$$

For a 1.8 V magnitude sinusoid this expression implies that $V_{tr(\text{max})} = 2.825$ V. At the first breakpoint $v_o = v_{tr} = \pm0.6$ V

$$m_2 = 0.945 \quad\Rightarrow\quad R_1 = 17.17 R_s.$$

At the second breakpoint $v_o = 1.2$ V, $v_{tr} = 1.27$ V

$$m_3 = 0.7613 \quad\Rightarrow\quad R_2 = 2.690 R_s.$$

Reasonable choices for the resistors, using standard values are:

$$R_s = 1\,\text{k}\Omega \quad R_1 = 17.2\,\text{k}\Omega \quad\text{and}\quad R_2 = 2.7\,\text{k}\Omega.$$

A PSpice analysis of the circuit was performed: The input and output waveforms are shown. Total harmonic distortion of the near-sinusoidal output of this circuit was calculated to be about 1.0%: a reasonable value for such a simple circuit. Typical converters usually have at least 6 breakpoints on each side of ground.

---

[6]These equations assume the forward dynamic resistance of the diodes is zero-valued. If the dynamic resistance can be reasonably approximated, the resistor values, R_1 and R_2, should be reduced by that approximation.

```
Triangle-tosinusoid converter
VS 1 0 PWL(0 -2.825 1M 2.825 2M -2.825)
RS 1 2 1K
D1P 2 10 D1N4148
D1N 10 2 D1N4148
R1  10 0  17.2K
D2P1 2  21 D1N4148
D2N1 22 2 D1N4148
D2P2 21 20 D1N4148
D2N2 20 22 D1N4148
R2  20 0  2.7K
D3P1 2 31 D1N4148
D3N1 33 2 D1N4148
D3P2 31 32 D1N4148
D3N2 34 33 D1N4148
D3P3 32 0 D1N4148
D3N3 0 34 D1N4148
.TRAN .01M 2M 0 0.01M
.LIB NOM.LIB
.PROBE
.FOUR 500 V(2)
.END
```

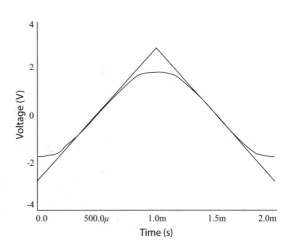

Another similar form of piece-wise linear two-port network uses BJTs as the switching device rather than diodes. A seven-segment array demonstrating this technique applied to triangle to sinusoid conversion is shown in Figure 13.16.[7] This design was based on the breakpoints and slopes derived in Design Example 13.5. The BJT-based circuit has several advantages over the diode-based circuit. Primary among the advantages is the flexibility concerning the amplitude of v_{tr}. As long as

$$V_{CC} \approx V_{tr(\max)} \quad \text{and} \quad 1.8 < V_{tr(\max)} < 15\,\text{V},$$

this circuits performs well with less than 1.5% total harmonic distortion. One particular drawback is the rather strong dependence on the forward-biased dynamic resistance of the base-emitter junctions of the various transistors. Performance is greatly dependent on these values (in the design shown, the 2.7 kΩ resistor was reduced to 2 kΩ and the 17.2 kΩ resistor was reduced to 14.3 kΩ due to dynamic resistance considerations).

Another technique useful in the conversion of triangular waves to sinusoids is non-linear amplification. A particularly useful circuit that employs this technique is shown in Figure 13.17. The design utilizes logarithmic amplification obtained with an overdriven differential gain stage: input triangle waves alternately force one the two BJTs to the verge of saturation.

In this amplifier the triangle wave is amplified linearly near the zero crossing: in the regions near the peaks of triangle the amplification is logarithmic. For a well-controlled triangle voltage input,

[7]More complete discussion can be found in Grebene, 1984, pp 592–595.

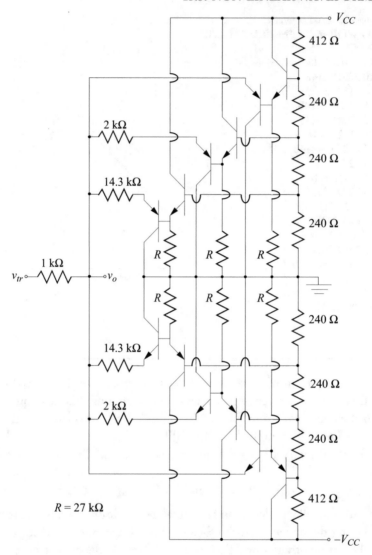

Figure 13.16: A seven-segment triangle to sinusoid converter.

the resultant transfer function is near-ideal and produces a low-distortion sinusoid output, v_o. Optimal performance occurs with

$$52\,\text{mV} < I_1 R < 86\,\text{mV},$$

and

$$V_{tr(\text{max})} \approx 95\,\text{mV}.$$

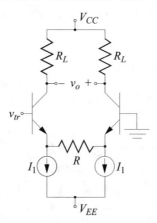

Figure 13.17: Differential pair triangle-sinusoid non-linear amplifier.

A logarithmic triangle-sinusoid converter is particularly useful at high frequencies where the higher-voltage input requirement of piece-wise linear converters may be difficult to produce with sufficient accuracy.

13.4   INTEGRATED CIRCUIT MULTIVIBRATORS

An alternative to the realization of multivibrators with discrete elements exists in the form of an integrated circuit package. These IC packages, commonly identified as *IC timers*, contain the primary components needed to implement monostable and astable multivibrators: a minimum number of external components is necessary to complete the multivibrator design. Among the IC timers currently available, the 555 timer has gained widest acceptance due to its versatility and low cost.

The basic functional block diagram of a 555 timer is shown in Figure 13.18. It consists of two comparators, a resistive network that sets the trigger levels of the comparators, a transistor that acts as a switch, and a set-reset (SR) flip-flop. The three internal resistors, labeled R, divide the input voltage, V_{CC}, so that the voltage trigger levels of the two comparators, CP_1 and CP_2, are at $2/3\,V_{CC}$ and $1/3\,V_{CC}$, respectively. These trigger levels can be altered by applying an external voltage at the *Control* input terminal. The outputs of the comparators control the state of the SR flip-flop. An SR flip-flop is a form of bistable circuit:[8] it is a level retention circuit with complementary outputs Q and $\bar{Q}$. The output, Q, transitions to match the S input when only one input is HIGH. When both inputs are LOW the flip-flop retains its last value of Q and holds it until at least one input transitions to HIGH. The terms *set* and *reset* refer to the action of the output, Q. Q *sets* (transitions to HIGH) when the set input, S, is HIGH: Q *resets* (transitions

[8]More information on the characteristics of flip-flops can be found in Section 16.5.

to LOW) when the reset input, R, is HIGH. The transistor switch shorts the *Discharge* input to ground when $\bar{Q}$ is HIGH.

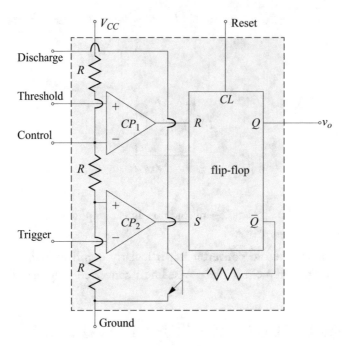

Figure 13.18: Simplified functional diagram of 555 IC timer.

The 555 timer is available in both bipolar and CMOS technologies. In various forms it is capable of producing timing signals with a duration that ranges from microseconds to hours. Astable oscillation up to a few Megahertz is possible. While many functional operations can be performed with this circuit, discussion will be limited to a monostable multivibrator (one-shot) and an astable multivibrator (non-linear oscillator).

13.4.1 A 555 TIMER MONOSTABLE MULTIVIBRATOR

Timing operations, using a 555 timer, are typically based on the charging and discharging of external networks. Most fundamental of these timing operations is that of a monostable multi-vibrator. The 555 timer implementation of a monostable multivibrator is shown in Figure 13.19. Here the timed interval is controlled by a single external RC network. The basic connections are as follows:

- A capacitor, C, is connected from the *threshold* input to *ground*.

- A resistor, R_A, is connected from the *threshold* input to the *positive power*, V_{CC}.

- The *discharge* and *threshold* inputs are shorted together.

- The *reset* input is held HIGH.

- The *control* input is left open (or connected through a small capacitor to ground).

The stable state of this circuit exists when the output, Q, of the flip-flop is LOW. In order to achieve that state the input, v_{in}, must be greater than the trigger level of CP_2. In this state, the complementary output of the flip-flop, $\bar{Q}$ is HIGH. This causes the transistor switch to activate, forcing a rapid discharge of the capacitor voltage to essentially zero. The output of CP_1 is LOW. The output of each comparator is therefore LOW: the LOW inputs to the flip-flop retain the LOW output until an input change occurs.

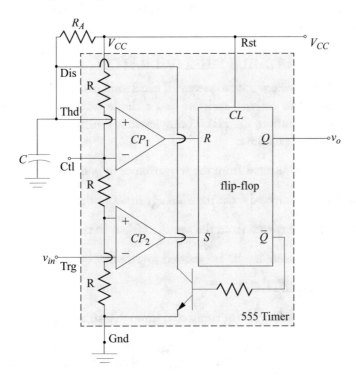

Figure 13.19: A 555 timer monostable multivibrator.

The quasi-stable state occurs when the input voltage, v_{in}, momentarily drops below the trigger level of CP_2:

$$v_{in} < V_{CC}/3. \tag{13.33}$$

This change in input level forces the output of CP_2 to a HIGH state, setting the flip-flop (Q, HIGH and $\bar{Q}$ LOW). The $\bar{Q}$ output of the flip-flop deactivates the transistor switch and allows the capacitor to begin charging toward the positive power voltage, V_{CC}:

$$v_c(t) = V_{CC}\left(1 - e^{-\frac{t}{R_A C}}\right). \tag{13.34}$$

The charging will continue until $v_c(t)$ exceeds the trigger voltage of CP_1. The duration of the charging is given by the solution to:

$$\frac{2V_{CC}}{3} = V_{CC}\left(1 - e^{-\frac{t}{R_A C}}\right) \quad \Rightarrow \quad t = R_A C \ln(3). \tag{13.35}$$

When $v_c(t)$ reaches the trigger level the output of CP_1 goes HIGH. If the input signal $v_i(t)$ has returned HIGH, the flip-flop changes state and returns to $Q = $ LOW: a single HIGH pulse of duration $R_A C \ln(3)$ is formed. If $v_i(t)$ is still LOW, the HIGH pulse continues until $v_i(t)$ goes HIGH at which time the pulse terminates.

13.4.2 A 555 TIMER ASTABLE MULTIVIBRATOR

An astable multivibrator where the frequency of oscillation and the duty cycle can be independently controlled with two external resistors and a single external capacitor is shown in Figure 13.20. In this configuration, the HIGH duty cycle is limited to the range of 50% to 100%. The basic connections are as follows:

- A capacitor, C, is connected from the *trigger* input to *ground.*

- A resistor, R_A, is connected from the *discharge* input to the *positive power*, V_{CC}.

- A resistor, R_B, is connected from the *threshold* input to the *discharge* input.

- The *trigger* and *threshold* inputs are shorted together.

- The *reset* input is held HIGH.

- The *control* input is left open (or connected through a small capacitor to ground).

In this astable configuration, the capacitor voltage, $v_c(t)$, transitions exponentially between the trigger levels of the two comparators (established by the resistors labeled R):

$$\frac{V_{CC}}{3} \leq v_c(t) \leq \frac{2\,V_{CC}}{3}. \tag{13.36}$$

Both the direction and rate of capacitor voltage transition are controlled by the switching transistor. When $v_c(t)$ is *between* the trigger levels, both comparators will have a LOW output: this output state signals the flip-flop to retain its last output state. When $v_c(t)$ reaches the upper trigger level, the output of CP_1 momentarily changes to a HIGH: this action activates the reset state of the flip-flop (Q, LOW and, $\bar{Q}$ HIGH). The output of the flip-flop activates the transistor

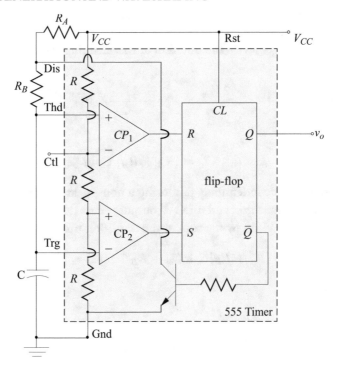

Figure 13.20: A 555 timer astable multivibrator.

switch which forces the capacitor to begin an exponential discharge through the resistor, R_B, to ground:

$$v_c(t) = \frac{2V_{CC}}{3} e^{-\frac{t}{R_B C}}. \tag{13.37}$$

As soon as the voltage drops below the upper trigger level, the output of CP_1 returns LOW, however, the flip-flop retains its output state and the discharge continues. It continues until $v_c(t)$ reaches the lower trigger level. The duration of this discharging transition, t_d, occurs at the solution to:

$$\frac{V_{CC}}{3} = \frac{2V_{CC}}{3} e^{-\frac{t_d}{R_B C}} \quad \Rightarrow \quad t_d = R_B C \ln(2). \tag{13.38}$$

When $v_c(t)$ reaches the lower trigger level, the output of CP_2 momentarily changes to a HIGH: this action activates the set state of the flip-flop (Q, HIGH and $\bar{Q}$ LOW). The output of the flip-flop deactivates the transistor switch which forces the capacitor to begin exponential charging through the resistors, R_A and R_B, toward V_{CC}:

$$v_c(t) = V_{CC} - \frac{2V_{CC}}{3} e^{-\frac{t}{[R_A + R_B] C}}. \tag{13.39}$$

As soon as the voltage rises above the lower trigger level, the output of CP_2 returns LOW, however, the flip-flop retains its output state and the charging continues. It continues until $v_c(t)$ reaches the upper trigger level. The duration of this charging transition, t_c, occurs at the solution to:

$$\frac{2\,V_{CC}}{3} = V_{CC} - \frac{2\,V_{CC}}{3}e^{-\frac{t_c}{[R_A+R_B]C}},$$

(13.40)

or

$$t_c = [R_A + R_B]\,C\,\ln(2).$$

(13.41)

The transitions continue indefinitely producing a non-linear oscillation. A HIGH output occurs during the charging transition and a LOW output occurs during the discharging transition. The period of oscillation is given by the sum of the transition times:

$$\tau = t_c + t_d = [R_A + R_B]C\,\ln(2) + R_B C\,\ln(2) = [R_A + 2R_B]C\,\ln(2),$$

(13.42)

or

$$\tau \approx 0.693[R_A + 2R_B]C.$$

(13.43)

The HIGH duty cycle is given by the ratio of t_c to τ:

$$\text{duty cycle} = \frac{t_c}{\tau} = \frac{R_A + R_B}{R_A + 2\,R_B}.$$

(13.44)

Since it is not possible to set $R_A = 0$ and still have a functioning astable multivibrator, the HIGH duty cycle will always be greater than 0.5 (50%). The output voltage and capacitor voltage waveforms for a 60% duty cycle 555 timer astable multivibrator are shown in Figure 13.21.

13.5 CONCLUDING REMARKS

A variety of circuits that generate non-sinusoidal waveforms has been presented in this chapter. These circuits are based on electronic multivibrators. Multivibrators are classified into three basic types based on the stability of the output states:

- Bistable Multivibrator—two stable output states.

- Astable Multivibrator—two quasi-stable output states.

- Monostable Multivibrator—one stable and one quasi-stable output state.

Bistable multivibrators provide two distinct output states when appropriately triggered: one of the most useful of these circuits is the Schmitt trigger. A Schmitt trigger performs the particularly useful task of noise elimination through hysteresis in its transfer function. Astable multivibrators

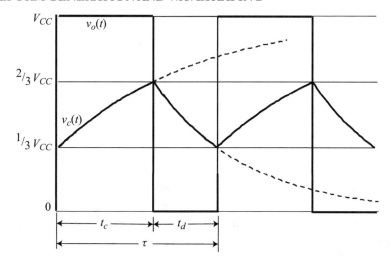

Figure 13.21: Waveforms for 555 timer astable multivibrator (60% duty cycle).

are non-linear oscillators that provide periodic square and/or triangular waveforms. Monostable multivibrators provide a single output pulse of fixed duration when triggered. Monostable and astable multivibrators can be realized through the use of an IC timer. Most dominant among IC timers is the 555 timer family.

Arbitrary, periodic waveforms can be derived from triangular waveforms. A technique to generate arbitrary waveforms using diode clipping circuitry is commonly used. Low harmonic content sinusoids can be generated using this technique or by non-linear amplification.

Summary Design Example

The transmission of digital computer signals over standard telephone lines requires the use of a modulator-demodulator unit known as a modem. Typically, modems transmit 2400 digital pulses per second. In the construction of such a device, a local square-wave oscillator is often required at that frequency. In order to interface with other digital circuits, the output of the oscillator must be TTL compatible. These design goals reduce to the following specifications:

$$f = 2400\,\mathrm{Hz} \qquad V_H \approx 5\,\mathrm{V} \qquad V_L \approx 0\,\mathrm{V}.$$

Design such a device so that the square-wave duty cycle lies between 40% and 60%.

Solution:

The obvious design alternatives are:

- a linear oscillator,

- a discrete-element astable multivibrator, or

- an IC timer astable multivibrator.

The two astable multivibrator types have a distinct advantage in complexity, size, and cost over a linear oscillator at this low frequency. In addition the TTL compatible output voltage levels lead to a 555 timer realization as an extremely advantageous choice. Therefore, the 555 timer circuit topology shown in Figure 13.20 is chosen as the basis for this design.

The specifications lead to specific parameter values needed in this design. If $V_{CC} = 5\,\text{V}$, the output voltage levels will meet specification. The frequency of oscillation requirement leads to:

$$[R_A + 2R_B]C \ \ln(2) = \frac{1}{2400} = 416.7\,\mu\text{s}.$$

Arbitrarily choose the capacitor to be a convenient standard value:

$$C = 0.027\,\mu\text{F} \quad \Rightarrow \quad R_A + 2R_B = 22.26\,\text{k}\Omega.$$

A 555 timer can only have duty cycle greater than 50%. Arbitrarily choose the duty cycle to lie easily within the design goals at 55%. This choice leads to a ratio of resistor values:

$$R_A + R_B = 0.55[R_A + 2R_B] \quad \Rightarrow \quad R_B = 4.5R_A.$$

Combining the two constraining equations for resistance values leads to standard-value resistors of value:

$$R_A = 2.23\,\text{k}\Omega \quad \text{and} \quad R_B = 10.0\,\text{k}\Omega.$$

The design is complete. If exact-value components are used the computed frequency is 2403.7 Hz (0.15% error) with a duty cycle of 55.02% (within specifications).

Design Verification Using SPICE

Most versions of SPICE are supplied with a macromodel of a CMOS version of the 555 timer. The simple astable multivibrator oscillator in this design example can be modeled as:

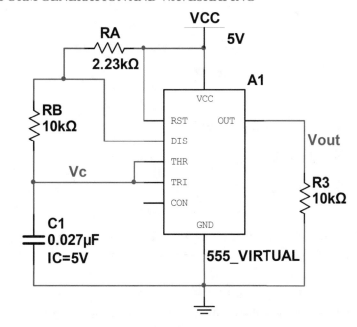

A plot of the output and capacitor voltages is shown below. The simulation yields a square wave of frequency 2378 Hz (−0.92% error) with 54.5% duty cycle (within specifications).

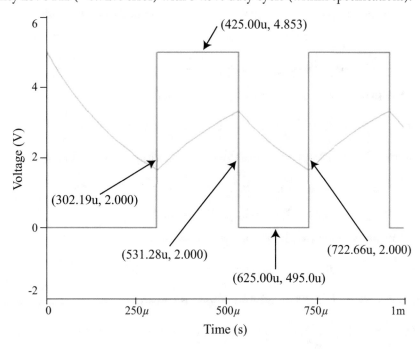

13.6   PROBLEMS

13.1.   A simple Schmitt trigger has the following design requirements:

- stable output voltages at ± 5 V.
- reference voltage, $v_r = 0$ V.
- threshold voltages, $V_T^+ = \pm 1$ V.

Design a circuit to meet these requirements to within ± 0.05 V.

13.2.   A sinusoidal input voltage
$$v(t) = A \sin(1000\,\pi t),$$

is the input to the simple Schmitt trigger circuit created in Design Example 13.1. Quantitatively describe the output voltage if:

(a) The input sinusoidal amplitude is 1.0 V (i.e., $A = 1.0$).

(b) The input sinusoidal amplitude is 0.4 V (i.e., $A = 0.4$).

Assume the output voltage is zero-valued at $t = 0$.

13.3.   A non-inverting Schmitt trigger has the following design requirements:

- stable output voltage at ± 10 V.
- positive slope threshold voltage, $V_T^+ = 2$ V.
- negative slope threshold voltage, $V_T^- = 0$ V.

Design a circuit to meet these requirements to within ± 0.05 V.

13.4.   An inverting Schmitt trigger has the following design requirements:

- stable output voltage at ± 10 V.
- positive slope threshold voltage, $V_T^+ = 1.5$ V.
- negative slope threshold voltage, $V_T^- = -3$ V.

Design a circuit to meet these requirements to within ± 0.02 V.

13.5.   Another form of OpAmp Schmitt trigger is shown.

(a) Determine the transfer characteristic as a function of the circuit parameters. Assume $V_{ref} < V_{CC} - 2$.

(b) If the circuit is constructed with the following circuit element values, what are the threshold and output voltages?

$$R_g = 8.2\,\text{k}\Omega \quad R = 1\,\text{k}\Omega$$
$$V_{ref} = 4\,\text{V} \quad\quad V_{CC} = 15\,\text{V}.$$

(c) Use SPICE to verify the results of part b).

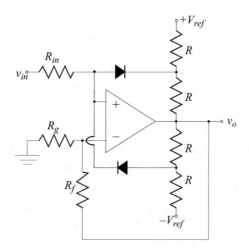

13.6. Design an astable multivibrator to produce a symmetric square wave of frequency 2 kHz. Verify the design using SPICE.

13.7. Design an astable multivibrator to produce a symmetric square wave with amplitude, $6\,V_{p-p}$ and frequency, 1.2 kHz. Verify the design using SPICE.

13.8. The astable multivibrator circuit shown is constructed with the following component properties:

$$R_f = 2\,\text{k}\Omega \quad\quad D_1 : V_\gamma = 0.6\,\text{V};\ V_z = 6.4\,\text{V}$$
$$R_g = 1\,\text{k}\Omega \quad\quad D_2 : V_\gamma = 0.6\,\text{V};\ V_z = 3.4\,\text{V}.$$

(a) What are V_H and V_L?

(b) Determine the HIGH duty cycle (t_{HL}/τ).

(c) Complete the design by choosing component values R and C so that the circuit will oscillate at 1 kHz.

(d) Verify the results using SPICE.

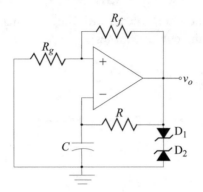

13.9. Design an astable multivibrator to produce a square wave of frequency 500 Hz with a HIGH state duty cycle of 40%. It is required that $V_H = 5$ V: V_L may be varied. Verify the design using SPICE.

13.10. Design an astable multivibrator to produce a square wave of frequency 500 Hz with a HIGH state duty cycle of 40%. It is required that $V_H = -V_L = 5$ V. Hint: In order to have different time constants for the two transitions, diodes may be placed in the discharge path. Verify the design using SPICE.

13.11. Design a one-shot circuit that responds to a pulse input with an output pulse of duration 3 ms. Verify the design using SPICE (any OpAmp macromodel will suffice in modeling a comparator in this application).

13.12. Design a one-shot circuit that responds to a pulse input with an output pulse of duration 1.0 ms. Verify the design using SPICE (any OpAmp macromodel will suffice in modeling a comparator in this application).

13.13. A stable 120 Hz pulse train of amplitude 5 V is required. Since commercial power is extremely stable, it has been decided to use the 60 Hz, 110 V power-line voltage as a triggering source for this pulse train. Design a system that will produce the required pulse train using a simple one-shot based on the simple design topology shown. Verify the design using SPICE.

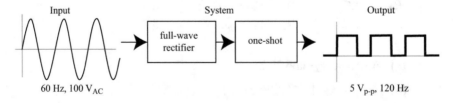

13.14. A stable 60 Hz pulse train of amplitude 10 V is required. Using commercial power-line input and a design topology similar to that in the previous problem, design a system to produce the required pulse train. Verify the design using SPICE.

13.15. Design an astable multivibrator to produce a 340 Hz symmetric triangle wave with peak-to-peak amplitude of 8 V. Verify the design using SPICE.

13.16. Design an astable multivibrator to produce a 500 Hz triangle wave with an amplitude of 10 V peak-to-peak and a 40% positive slope duty cycle. Verify the design using SPICE.

13.17. It is suggested that a device to sound the orchestral tuning note ($A = 440\,\mathrm{Hz}$) could be inexpensively mass-produced using an astable multivibrator as the tone oscillator. The presence of harmonics of the 440 Hz tone in the output is desirable.

  (a) Design an astable multivibrator to produce that frequency.

  (b) Trained musicians are easily capable of determining pitch (frequency) error of ± 3 cents. A cent is defined as 1% of a semitone, where a pitch transition of a semitone produces a change in frequency by a factor of the twelfth root of two. Thus, pitches (when compared to the ideal pitch frequency, f_o) that are considered "in tune" must lie within the frequency range:

$$\left(\sqrt[12]{2}\right)^{\frac{-3}{100}} f_o < f < \left(\sqrt[12]{2}\right)^{\frac{3}{100}} f_o \quad \Rightarrow \quad 2^{\frac{-3}{1200}} f_o < f < 2^{\frac{3}{1200}} f_o.$$

  Comment on the practicality of using the design of part a) for this mass-produced device.

13.18. Design a voltage controlled oscillator that will output a 10 V peak-to-peak square wave of variable frequency from 50 Hz to 1 kHz for an input voltage range of 0.2 V to 4 V.

13.19. Design a voltage controlled oscillator that will output a 10 V peak-to-peak triangle wave of variable frequency from 50 Hz to 4 kHz for an input voltage range of 0.1 V to 8 V.

13.20. The diode network shown is a proposed triangle-to-sinusoid converter. It is said to be capable of taking a 20 V_{p-p} triangle-wave input and produce a sinusoid with less than 3% total harmonic distortion. Unfortunately the design does not include the proper value for the reference voltages $\pm V_{ref}$.

  (a) Determine the break point voltages as a function of V_{ref} (assume $V_{ref} > V_\gamma$).

  (b) Choose an appropriate value for V_{ref} and determine the total harmonic distortion of the output using SPICE.

  (c) Try another value for V_{ref}. Compare the resultant output total harmonic distortion with that of part b).

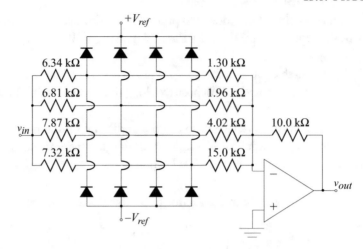

13.21. Design, using the basic topology shown in Figure 13.15 (some diodes must be reversed), a seven-segment expanding wave-shaping network for use in the receiver of the digital telephony system described above. The desired design characteristics are:

$$v_o = v_{in}, \qquad |v_{in}| < 1$$
$$v_o = 2\,v_{in} - 1, \qquad 1 < v_{in} < 2 \qquad v_o = 2\,v_{in} + 1, \qquad -1 > v_{in} > -2$$
$$v_o = 4\,v_{in} - 5, \qquad 2 < v_{in} < 3 \qquad v_o = 4\,v_{in} + 5, \qquad -2 > v_{in} > -3$$
$$v_o = 8\,v_{in} - 17, \qquad 3 < v_{in} < 4 \qquad v_o = 8\,v_{in} + 17, \qquad -3 > v_{in} > -4.$$

Hint: An amplifier is needed in addition to the passive diode network.

13.22. One common use of non-linear waveshaping occurs in digital telephony. Analog-to-digital (A/D) conversion introduces the same quantization noise for small-amplitude signals than for large-amplitude signals. Thus, small signals appear to have more *relative* noise due to the A/D conversion than large signals. In order to equalize the signal-to-noise ratio, the input message is amplified in a non-linear fashion prior to A/D conversion: this process is called *compressing*. Upon receipt, the message is digital-to-analog (D/A) converted and non-linearly amplified to restore linearity, this process is called *expanding*. Together the two processes are called *companding*. Design, using the basic topology shown in Figure 13.15, a seven-segment compressing wave-shaping network with the following characteristics:

$$v_o = v_{in}, \qquad\qquad |v_{in}| < 1$$
$$v_o = 0.5\,(v_{in} + 1), \qquad 1 < v_{in} < 3$$
$$v_o = 0.25\,(v_{in} + 5), \qquad 3 < v_{in} < 7$$
$$v_o = 0.125\,(v_{in} + 17), \quad 7 < v_{in} < 15$$
$$v_o = 0.5\,(v_{in} - 1), \qquad -1 > v_{in} > -3$$
$$v_o = 0.25\,(v_{in} - 5), \qquad -3 > v_{in} > -7$$
$$v_o = 0.125\,(v_{in} - 17), \quad -7 > v_{in} > -15.$$

13.23. Design a monostable multivibrator using a 555 timer with a pulse output of duration 10 ms. The input is a 5 V signal that drops to 0 V for a duration of 1 ms to trigger the multivibrator.

13.24. Series connection of IC timers can produce an output consisting of a delayed pulse of fixed duration. The first timer fixes the delay, with its negative transition triggering the second timer. The second timer sets the duration of the output pulse. The two timers are capacitively coupled with a pull-up resistor on the input of the second timer: the RC time constant of this coupling circuit must typically be less than 50 μs (for a 555 timer). Design such a two IC timer circuit that will produce an output pulse that is delayed from an input triggering pulse. The design specifications are:

Input trigger   — 5 V to 0 V for 10 μs
Output delay   — 2 ms $\pm$ 10 μs
Output duration   — 0.5 ms $\pm$ 10 μs.

Use SPICE to verify the design.

13.25. Verify the waveforms shown in Figure 13.21 by designing a 555 timer astable multivibrator to oscillate at 2 kHz with 60% HIGH duty cycle. Use SPICE to display the waveforms.

13.26. The expression for the output-waveform duty cycle of an IC timer oscillator is given by Equation (13.44). A 50% duty cycle implies that $R_A = 0$ (the *discharge* and V_{CC} terminals are shorted together). Explain why the circuit will not oscillate with a 50% duty cycle. Use SPICE simulation to verify that the circuit will not oscillate.

13.27. The circuit shown purports to use an IC timer to produce square-wave oscillation with a 50% duty cycle. The design parameters are:

$$f = \frac{1}{1.386\,R_1 C_1} \quad \text{and} \quad R_2 > 10R_1.$$

(a) Use this circuit topology to design a 540 Hz oscillator.

(b) Use SPICE and the macromodel of a 555 timer to check the design for frequency and duty cycle compliance with specifications.

(c) If design parameters are not met, explain any mechanisms that cause the variation.

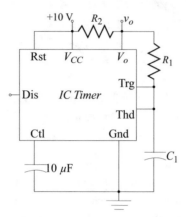

13.28. The circuit shown is a voltage-to-frequency converter based upon linear charging of a capacitor by a constant current source. Use SPICE and the macromodels for a 741 OpAmp and 555 timer to determine the linearity of the voltage-to-frequency conversion over the range $0.5\,\text{V} < v_{in} < 5\,\text{V}$:

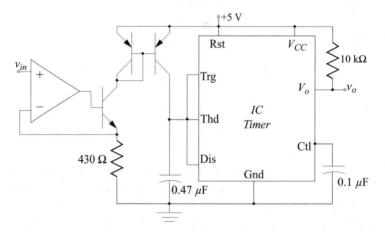

13.29. One-shot circuits that use an IC timer have a very rapid recovery time. This rapid recovery creates opportunities for unusual applications of the timer. One application is the use of an IC timer monostable multivibrator (one-shot) to divide the frequency of a pulse train. In order to divide the frequency by a factor, N, the pulse duration of the of the one-shot is chosen to lie in the range:

$$(N - DC)\tau < t < N\tau,$$

where τ is the period of the original pulse train and DC is the HIGH state duty cycle given as a fraction rather than a percentage. The trigger terminal serves as the input to the frequency divider. Design, using a 555 timer, a circuit that will divide the frequency of a 10 kHz square wave ($DC = 0.5$) by a factor of three ($N = 3$). Verify the design using SPICE.

13.30. The circuit shown may be used to detect irregularities in a train of pulses. As long as the pulse spacing is shorter than the timing interval of the timing circuit, the monostable circuit is continuously triggered. Pulse spacing greater than the timing interval or the termination of the pulse train allows completion of the timing interval and the generation of an output pulse.

    (a) Assume a pulse train input at 1 kHz with a LOW duty cycle of 20%. Complete the design so that a missing LOW pulse will be detected (the pulse spacing is effectively increased by the absence of a pulse).

    (b) Test the design of part a using SPICE.

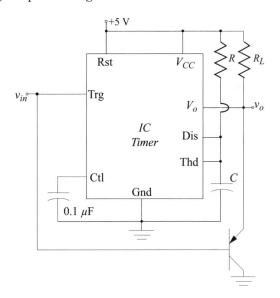

13.7   REFERENCES

Linear Circuits Data Book, Volume 3, Voltage Regulators/Supervisors, Comparators, Special Functions, and Building Blocks, Texas Instruments Inc., Dallas, 1992.

    Grebene, Alan B., *Bipolar and MOS Analog Integrated Circuit Design,* John Wiley & Sons, New York, 1984.

    Hambley, Allan R., *Electronics, A Top-Down Approach to Computer-Aided Circuit Design,* Macmillan Publishing Company, New York, 1994.

Millman, Jacob, *Microelectronics, Digital and Analog Circuits and Systems,* McGraw-Hill Book Company, New York, 1979.

Millman, Jacob and Taub, Herbert, *Pulse, Digital, and Switching Waveforms,* McGraw-Hill Book Company, New York, 1965.

Savant, C.J., Roden, Martin S., and Carpenter, Gordon L., *Electronic Circuit Design, An Engineering Approach,* The Benjamin/Cummings Publishing Co., Menlo Park, 1987.

Sedra, Adel S. and Smith, Kenneth C., *Microelectronic Circuits,* Saunders College Publishing, Phil.

Wojslaw, Charles F. and Moustakas, Evangelos A., *Operational Amplifiers*, John Wiley & Sons, New York, 1986.

CHAPTER 14

Power Circuits

All the electronic circuits described in the first three sections of this book have as their main purpose the modification of an input signal so as to perform useful work on the load. This signal modification may take the form of signal amplification, frequency filtering, digital logical operations, or a variety of other possibilities. Inherent in all designs is an electrical power source. Most often this source is a DC voltage source: in some instances, the DC is derived from AC source. The proper operation of all electronic circuitry depends on the application of uniform electrical power in the form of these sources. Power electronic circuits are responsible for converting the available electric energy into a form with appropriate uniformity.

Regardless of the form of the input electrical energy (AC, DC, or a combination) the basic form of an electrical power system remains consistent. A typical electrical power system is shown in Figure 14.1. In such a system the input energy is filtered, converted to a new form, or shifted to a new level by an electrical power circuit, and again filtered. Electronic (or possibly mechanical) observation of the load conditions is an important factor in control of all operations.

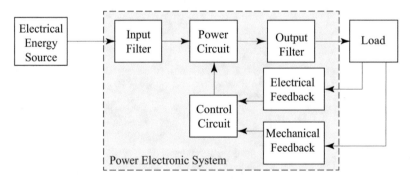

Figure 14.1: Typical electronic power system.

Previous chapters have described the principles of AC to DC conversion through the use of diode half-wave or full-wave rectifiers followed with a simple passive low-pass filter (active filtering is extremely inefficient in power applications). The benefits of feedback on stability have also been discussed. This chapter focuses on combining previously explored devices and principles into power circuits useful to the electronic designer. While many of these devices are commercially available in IC packages, knowledge of the principles of operation is valuable to the prudent device user or circuit designer.

Within many power circuits there exists the need for a high-power switch. BJTs and FETs are extremely useful as switching elements and have previously been discussed. Both of these transistor switches find common use in power circuits. Thyristors form another family of semi-conductor devices that are extremely useful in switching applications. Most common among this family are the Silicon controlled rectifier, typically used in DC applications, and the triac for AC applications.

Voltage regulators are a common device available for providing a stable DC output voltage for a wide range of input voltage and output current. As realized in a three-pin IC package, these regulators are extremely effective, easy to implement, and inexpensive. The internal design of both linear and switching regulators is discussed. While the design of DC power supplies is not specifically addressed in this chapter, many of the design principles of simple voltage regulators can be applied to their more powerful relative.

Protection against high-voltage transient or the accidental application of an overvoltage is another major concern of the electronic designer. Several types of transient suppressors are discussed. Similarly, several overvoltage protection circuits are described.

14.1   THYRISTORS

Thyristors are a form of solid state switch that is activated by a triggering signal. Prior to activation, a thyristor acts as an extremely high-impedance path: once activated, a thyristor acts as a low-impedance path and *remains activated* until the switched current falls below a minimum value, the holding level, at which time it deactivates. Once a thyristor is activated, the triggering signal is no longer necessary to continue activation unless low switched current deactivates the thyristor. The most common members of the thyristor family include:

- Silicon controlled rectifiers (SCRs)

- Triacs

- Programmable unijunction transistors (PUTs)

- Silicon bilateral switches (SBSs)

- Sidacs

Silicon controlled rectifiers are most commonly used as power control elements, triacs are bi-directional switches and are most useful in AC power applications under 40 Amperes, pro-grammable unijunction transistors are most often used in timer circuits, Silicon bilateral switches are most commonly used as gate trigger devices for the power control elements, and Sidacs are a high voltage bilateral trigger device. The thyristor family also includes the gate turnoff (GTO) thyristor: a device capable of being turned off, as well as on, with the application of different polarity gate current. The GTO thyristor typically is used in power inverters (DC to AC power converters).

Thyristors have several distinct advantages over mechanical switches. Primary among these advantages over mechanical switches are:

- High switching speed.

- Low energy switch triggering.

- Automatic debounce characteristics.

- Zero-current deactivation avoids contact arcing in inductive circuits.

While each type of thyristor has its particular uses, discussion in this chapter will be limited to the two most commonly used thyristor types, the SCR and the Triac. Several SCR applications will be presented in later sections of this chapter.

14.1.1 THE SILICON CONTROLLED RECTIFIER

The Silicon controlled rectifier (SCR) is the foundation element of the thyristor family. Its circuit symbol is shown in Figure 14.2. The three SCR terminals take their names from diode terminology and are identified as:

A – anode

K – cathode

G – gate

The SCR can conduct currents in excess of 50 Amps from the anode to the cathode and is capable of blocking voltages up to 800 V. The gate terminal is the input for the triggering signal.

Figure 14.2: The SCR circuit symbol.

SCRs have a four-layer structure of alternating p-type and n-type semiconductor material. Conceptually, this structure takes the form of Figure 14.3a. The structure functionally acts as two complementary BJTs connected as shown in Figure 14.3b.

Analysis of the SCR equivalent circuit leads to its modes of operation. The SCR has two fundamental operational states:

- An OFF state where only extremely small leakage currents pass from the anode to the cathode. In this state transistor Q_1 is cutoff and Q_2 is in the forward-active region, albeit with extremely low current flow.

- An ON state where current flows freely between anode and cathode. Here both transistors are in saturation.

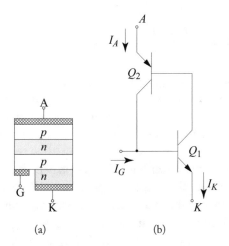

(a)   (b)

Figure 14.3: SCR structure (a) Typical medium-power SCR semiconductor structure; (b) Equivalent circuit.

In order to understand the transition between modes, it is necessary to examine the basic transistor current relationships:

$$I_{C1} = -\alpha_1 I_{E1} + I_{CO1} = \alpha_1 I_K + I_{CO1}, \tag{14.1}$$

and

$$I_{C2} = -\alpha_2 I_{E2} + I_{CO2} = -\alpha_2 I_A + I_{CO2}. \tag{14.2}$$

Here the two quantities I_{CO1} and I_{CO2} are the collector leakage currents of each transistor[1] (with an open emitter) and α_1 and α_2 are the collector-emitter current ratios. Note that in this formulation, α_i is transistor-region dependent and is only equal to α_{Fi} in the forward-active region. Kirchhoff's current law applied to the SCR equivalent circuit of Figure 14.3b yields expressions for the currents into the anode and out of the cathode of the SCR:

$$I_A = I_{C1} - I_{C2}, \tag{14.3}$$

and

$$I_K = I_A + I_G = I_{C1} - I_{C2} + I_G. \tag{14.4}$$

The above four equations can be combined to determine the SCR anode current as a function of the transistor parameters and the gate current:

$$I_A = \{\alpha_1 (I_A + I_G) + I_{CO1}\} - \{-\alpha_2 I_A + I_{CO2}\}, \tag{14.5}$$

[1] Since Q_1 is an *npn* BJT, the collector leakage current, I_{CO1}, is a positive quantity. Q_2 is a *pnp* BJT and has negative I_{CO2}.

which leads to

$$I_A = \frac{\alpha_1 I_G + I_{CO1} - I_{CO2}}{1 - (\alpha_1 + \alpha_2)}. \tag{14.6}$$

When the control transistor, Q_1, is OFF, $\alpha_1 \approx 0$ and $\alpha_2 < 1$: the SCR anode current is a small multiple of the sum of the magnitude of the leakage currents. As current is applied to the gate of the SCR and Q_1 begins to turn on, α_1 increases until the denominator of Equation (14.6) becomes zero. This singularity in the expression for the anode current, I_A, is physically realized by a rapid increase in the anode current until it is limited by the external circuitry to which the SCR is connected. At that current limiting point, I_G may be removed and the two BJT leakage currents will suffice to latch the SCR in the ON state. SCR deactivation will only occur when the anode current drops to zero value due to external circuit circumstances. SCR reactivation will occur when V_{AK} is again positive and a current pulse enters the gate. SCR condution current for the case of sinusoidal V_{AK} and a pulse train at the SCR gate is shown in Figure 14.4.

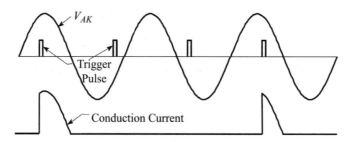

Figure 14.4: SCR conduction triggered by a pulse train.

While a conducting SCR presents very low dynamic resistance to the conduction current, there is a voltage drop across the anode-cathode terminals. The voltage across the SCR in the ON state is approximately given by the sum of a base-emitter drop, a collector-emitter drop and an ohmic loss:

$$V_{AK} \approx V_{BE(sat)} + V_{CE(sat)} + R_B I_A.$$

Here R_B is the bulk resistance associated with the SCR. The ON voltage drop has a minimum value in the range of 1 V and may be as high as 2–3 V for high-current SCRs. The actual voltage drop for any particular SCR is best determined from manufacturer's specifications or experimental evaluation.

In addition to the desired SCR activation with positive gate current, there are four false SCR activation mechanisms which must be avoided:

- High rate of change of anode-cathode voltage, dV_{AK}/dt.

- High anode-cathode voltage, V_{AK}.

- High device junction temperature.

- Energy injection into device semiconductor junctions, principally in an optical fashion.

Reasonable design caution can avoid false SCR activation by the last three mechanisms: opaque packaging avoids optical activation, a proper heat sink avoids high temperatures, and selection of an SCR with a correct voltage-blocking rating avoids an avalanche breakdown. Fast variation of the anode-cathode voltage, V_{AK}, is the most likely significant false activation mechanism in most applications.

A voltage applied across the anode-cathode terminals of an SCR induces a current through the SCR proportional to the derivative of the applied voltage. This current is due to the junction capacitances between the layers of semiconductor:[2]

$$i = C \frac{d V_{AK}}{dt}.$$

If this capacitance-charging current exceeds the gate triggering current, the SCR is subject to false triggering. Sensitivity to rate of change false activation can be reduced by a resistance shunting the gate-cathode junction or, more commonly, controlled with a snubber circuit.

Snubber circuits limit the rate of change of the voltage across a SCR (or more generally any thyristor). In their most simple form, a snubber circuit consists of a series connected resistor and capacitor which shunt the SCR, as shown in Figure 14.5. The RC time constant of the snubber circuit limits the rise time of the anode-cathode voltage and thereby reduces the possibility of false activation. On occasion the resistor in the snubber circuit may be shunted or connected in series with a diode. This diode is added to aid in suppression of transient voltages that may occur.

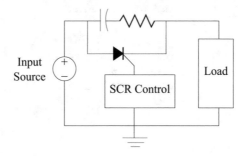

Figure 14.5: SCR application with snubber circuitry.

14.1.2 THE TRIAC

The triac is also a three-terminal semiconductor switch: its circuit symbol is shown in Figure 14.6. The triac switch differs from the SCR in that it is capable of conduction currents in either direc-

[2]Junction capacitance is discussed in Section 10.4 (Book 3).

tion. The current-carrying terminals are simply identified at MT_1 and MT_2 (Main Terminals 1 and 2): the triggering terminal, G, remains identified as the gate. Positive conduction current is identified as flowing from MT_2 through the triac to MT_1: positive gate current flows into the triac.

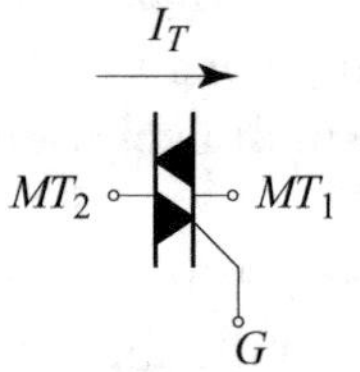

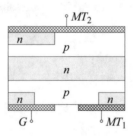

Figure 14.6: Triac circuit symbol.

The triac is a five-layer semiconductor device as shown in Figure 14.7. In many ways it can be thought of as two complementary SCRs (a *pnpn* device and an *npnp* device) connected in parallel. There are, however, properties of the triac that can not be described by such a simple model. In particular, is the ability of the triac to be triggered into conduction in *either direction* by either a positive or negative current into the gate. This bipolar triggering ability can be accounted for by noting the region between the gate and MT_1 is essentially two complementary diodes. Either a positive or negative gate current will bring one of these diodes into conduction triggering the same transistor action found in a SCR.

Figure 14.7: Typical triac semiconductor structure.

The four possible regions of triac operation are identified in Figure 14.8: the regions are identified by the polarity of the main terminal voltage difference and the polarity of the gate current. While it is possible to activate the triac in any region, the activation sensitivity to the gate current varies with region.

Region I (positive gate current and positive main terminal voltage difference) is the most sensitive of the four regions. Regions II and III are only slightly less sensitive to gate current turn on. Region IV typically requires as much as four times the gate activation current as Region I: this

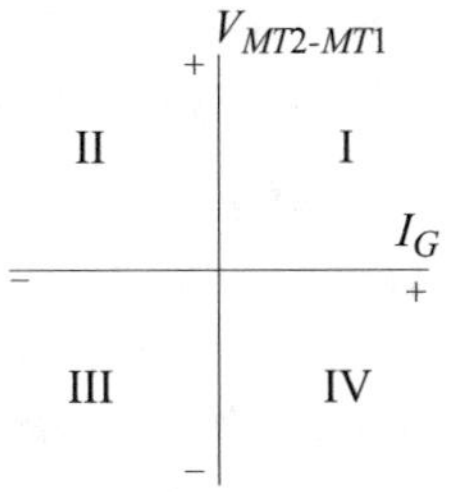

Figure 14.8: Triac regions of operation.

region of operation is avoided whenever possible. In many AC applications, the polarity of the gate current is automatically reversed as the polarity of the input voltage reverses so that operation is always in regions I and III.

The triac is subject to the same false activation mechanisms as the SCR. Many of the same precautions must be taken to ensure that the only activation mechanism is through the gate current. One additional precaution that must be taken involves the necessity of bidirectional snubber circuitry. While simple RC snubber circuitry is effective for both conduction current polarities (assuming proper choice of non-polarized capacitor), more complex SCR snubber circuitry may not be bidirectional and therefore inappropriate for triac use.

14.2   VOLTAGE REGULATOR DESIGN

The performance of electronic circuitry often depends on the application of stable DC power to the circuitry. Zener diode regulators, as discussed in Chapter 2 (Book 1), are one possible method for providing relatively stable DC power. However, Zener diode regulators have significant drawbacks: the most significant of which is the dependence the output voltage on load current and temperature. A better alternative to the use of a zener diode in efficiently ensuring stable power is the use of an integrated circuit voltage regulator. Such devices are highly effective, widely available, and relatively inexpensive. An integrated circuit voltage regulator provides specific, stable DC power over a wide range of load current and input voltage conditions and has relatively small variation in output with temperature.

An integrated circuit voltage regulator consists of three basic elements as shown in Figure 14.9:

- A voltage reference element that provides a known stable voltage level, V_{REF}, that is essentially independent of temperature and the input voltage.

- An error amplifier that compares the output voltage or some fraction of the output voltage to the reference voltage, V_{REF}.

- A power control element that converts, as indicated by the error amplifier, the input voltage to the desired output voltage over varying load conditions.

Each of these elements can be realized with several circuit topologies. The following discussion provides a sampling of the element topologies common in typical voltage regulators. While all examples shown here assume that the voltage to be regulated is positive, regulation of both voltage polarities is common and accomplished with the same general circuit topology. In addition, tracking regulators provide a symmetric pair of regulated voltages (i.e., $\pm$ 10 V) for applications that need matched, regulated power. One such application is the dual power bus needed for OpAmps.

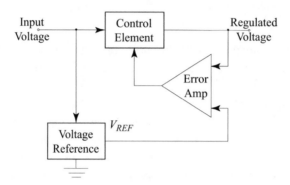

Figure 14.9: Typical voltage regulator block diagram.

14.2.1 THE VOLTAGE REFERENCE

The design goal of the voltage reference element is to provide a known, stable reference voltage, V_{REF}. There are three basic circuit topologies that provide a reference voltage within common voltage regulators. Two of these circuits are based on the Zener diode reverse breakdown voltage, V_z, and the third is based on the base-emitter junction voltage of a BJT. Simplest among the three circuits is a basic Zener diode reference, as shown in Figure 14.10.

In this circuit the output voltage, V_{REF}, is dependent on the diode zener breakdown voltage, V_z, the diode zener resistance, r_z, the input voltage, V_{in}, and the resistor value, R:

$$V_{REF} = V_z \left(\frac{R}{r_z + R} \right) + V_{in} \left(\frac{r_z}{r_z + R} \right). \tag{14.7}$$

The basic zener diode reference can be satisfactory in applications where the input voltage, V_{in}, is relatively stable. However, the susceptibility of this circuit to variation in input voltage and load-current induced temperature variation, may make it a poor choice in many applications.[3]

---

[3]The zener voltage for integrated circuit zener diodes varies with temperature by approximately $+2.2\,\text{mV}/^\circ\text{C}$.

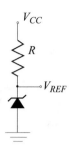

Figure 14.10: Basic Zener diode voltage reference.

An improvement in the basic zener diode voltage reference can be obtained by making the diode current independent of input voltage, V_{in}. One typical circuit topology that reduces diode current variation by driving the zener diode with a constant current source is shown in Figure 14.11.

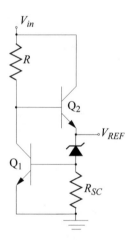

Figure 14.11: Constant-current Zener diode voltage reference.

Here the current in the zener diode is the sum of the base current in Q_1 and the current through resistor R_{SC}:

$$I_z = I_{B1} + I_{R_{SC}}. \tag{14.8}$$

In most applications the base current can be ignored:

$$I_z \approx I_{R_{SC}} = \frac{V_{BE1}}{R_{SC}}. \tag{14.9}$$

This gives a stable reference voltage, V_{REF}:

$$V_{REF} = V_z + I_z r_z + V_{BE1} \approx V_z + V_{BE1}\left(1 + \frac{r_z}{R_{SC}}\right). \tag{14.10}$$

Temperature variation of the reference voltage can be minimized by balancing the *positive* temperature variation of the zener voltage, V_z, with the *negative* temperature variation of the base-emitter voltage, V_{BE1}. The major disadvantage of the constant-current zener diode reference voltage is the need for the input voltage, V_{in}, to be relatively large: it must remain, depending on the exact circuit design, at least 1.5 volts more than the zener voltage.[4] This restriction on the input voltage limits the minimum voltage application of such a regulator.

A third common reference voltage circuit, the bandgap voltage reference, is shown in Figure 14.12. This design allows for a minimum input-output voltage difference as small as 0.6 volts: the input and output vary only by the voltage across the input resistor, R.

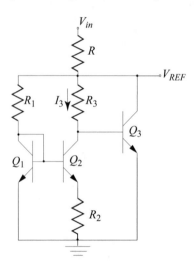

Figure 14.12: Bandgap voltage reference.

The output of the bandgap voltage reference is based on the highly predictable base-emitter voltage of a BJT in the forward-active region. Here the output reference voltage, V_{REF}, is given by:

$$V_{REF} = V_{BE3} + I_3 R_3. \tag{14.11}$$

The current, I_3, is essentially the output of a Widlar current source[5] formed by transistors Q_1 and Q_2 and resistors, R_1 and R_2. The current source has an output current, I_{C2}, given by the solution

---

[4]Manufacturers specify the minimum voltage difference between input and output voltage: for this design topology, it typically lies in the two to three volt range.
[5]Widlar current sources are discussed in Section 6.4.5 (Book 2).

to the transcendental equation

$$I_{C2}R_2 = \frac{\beta_F}{\beta_F + 1}\eta V_t \ln\left(\frac{I_{C1}}{I_{C2}}\right). \tag{14.12}$$

If the base current of Q_3 is assumed to be small compared to the collector current of Q_2, the output voltage can be derived from Equations (14.11) and (14.12):

$$V_{REF} \approx \eta V_t \frac{R_3}{R_2} \ln\left(\frac{I_{C1}}{I_{C2}}\right) + V_{BE3}. \tag{14.13}$$

It can be seen from Equation (14.13) that the bandgap reference circuit has an output, V_{REF}, that is independent of the input voltage, V_{in}. An additional benefit of this circuit topology is its relative insensitivity to temperature variation. If the circuit is built in IC form, then all circuit elements are essentially at the same temperature. The ratio of the collector currents I_{C1} and I_{C2} remains basically constant over temperature change and the variation in the reference voltage with temperature is given by:

$$\frac{\Delta V_{REF}}{\Delta T} \approx \eta \frac{R_3}{R_2} \ln\left(\frac{I_{C1}}{I_{C2}}\right) \frac{\Delta V_t}{\Delta T} + \frac{\Delta V_{BE3}}{\Delta T}. \tag{14.14}$$

In Chapter 2 (Book 1), the voltage equivalent temperature, V_t, was defined as:

$$V_t = \frac{kT}{q} \approx \frac{T}{11600}, \tag{14.15}$$

V_t consequently has a positive variation with temperature. In Chapter 3 (Book 1), it was shown that the variation in V_{BE} with temperature is negative. Thus, it is possible have essentially no temperature variation in V_{REF} if the two temperature dependent terms in Equation (14.14) cancel. The necessary conditions for cancellation are:

$$\frac{R_3}{R_2} \ln\left(\frac{I_{C1}}{I_{C2}}\right) = -\frac{11600}{\eta}\left(\frac{\Delta V_{BE3}}{\Delta T}\right). \tag{14.16}$$

A judicious choice of the resistors R_1, R_2 and R_3 allows the designer to produce a wide variety of reference voltages that are relatively insensitive to temperature and input voltage variation. A typical reference voltage value used in many regulators is 2.5 V.

All of the described voltage reference circuits find uses other than in voltage regulators: they can be used whenever an independent voltage reference is necessary. A precision reference voltage is usually achieved with either the bandgap or constant current zener reference circuits. These precision voltage reference circuits are used in wide variety of electronic applications. Most common among these is as reference for precision analog to digital conversion. A few of the common applications for voltage reference circuits are:

- Analog to digital conversion, A/D

- Digital to analog conversion, D/A

- Digital multimeter applications

- Voltage to frequency conversion, VCO

- Frequency to voltage conversion, FM detection

14.2.2 THE ERROR AMPLIFIER

The error amplifier in IC voltage regulators is basically a differential amplifier that compares the output voltage (or a fraction of the output voltage) to the reference voltage. They may take as simple a form as an emitter-coupled or source-coupled pair, but are more often similar to an OpAmp circuit.[6] These differential amplifiers must have a high common-mode rejection ratio (CMRR), low offset currents and voltages, and a high power supply rejection ratio (PSRR) to be effective in voltage regulation applications.

14.2.3 LINEAR VOLTAGE REGULATORS

The inputs to the power control element of a voltage regulator are the supply voltage and the control signal from the error amplifier. The output of this element must be a constant, specific voltage over a wide range of load currents and impedances. While the voltage reference and error amplifier elements can be virtually identical for a wide range of voltage regulators, the power control element varies widely depending on the type of desired regulation. The three basic types of power control elements are:

- Series

- Shunt

- Switching

Voltage regulators are often classified by the type of control unit employed. A voltage regulator with a series or shunt power control element is classified as a *linear voltage regulator*: a regulator with a switching power control element is a *switching regulator*. Linear regulators have a distinct noise advantage over switching regulators. Due to energy conversion efficiency considerations, linear regulators are typically found in low power electronic applications: switching regulators find greater use in high power applications where they act as regulated power supplies.

---

[6]Differential amplifiers are discussed in several sections of this book. See Chapters 2 (Book 1) and 6 (Book 2) for basic discussions.

Series Regulators

The series regulator is best suited for medium load current applications where the input-output voltage difference is not large. The commonly-used, three-terminal IC voltage regulator, found in many electronic designs, is usually of this design. Safety features such as input overvoltage and output short-circuit protection can also be provided in series regulators.

For a series regulator, the error amplifier regulates the output voltage, V_o, through an active series element, usually a transistor, as shown in Figure 14.13. If the output voltage falls below

$$V_o \leq \frac{R_2 + R_1}{R_2} \, V_{REF}, \tag{14.17}$$

the error amplifier will supply a positive voltage signal to the BJT. The transistor base-emitter junction becomes more forward biased and increases the current to the load. The output voltage is thereby increased until an appropriate balance is achieved. Large input-output voltage differences imply that the active control element must dissipate significant power. The internal losses lead to a typical working efficiency for a series regulator of 40–50%: high input-output voltage differential can reduce this figure further. The potentially large internal power loss is the greatest disadvantage of a series regulator.

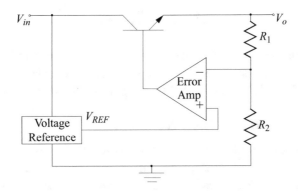

Figure 14.13: Basic series regulator topology.

Shunt Regulators

While the shunt regulator is usually the most inefficient of all regulator topologies, it can be a good choice in some applications. It is relatively insensitive to input voltage variation, protects the source from load current transients, and is inherently rugged against an accidental load short-circuit. It is the simplest of all regulators: for example, the zener diode regulator, as described in Chapter 2 (Book 1), is a passive form of shunt regulator. The basic circuit topology of an active shunt regulator is shown in Figure 14.14.

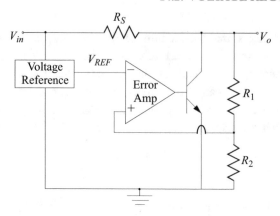

Figure 14.14: Basic shunt regulator topology.

Here, the active series pass element of the series regulator has been replaced by a resistor, R_S, and the changes in load current are neutralized by shunting excess current through an active element (usually a transistor) to ground. When the output voltage falls below

$$V_o \leq \frac{R_2 + R_1}{R_2} V_{REF}, \tag{14.18}$$

the error amplifier reduces voltage to the base of the BJT thereby reducing the shunt current. Since the current through the series pass element, R_S, must remain constant, more current is available for the load and V_o is increased until an appropriate balance is achieved.

14.2.4 SWITCHING VOLTAGE REGULATORS

The switching regulator is most often used in relatively high-power applications where power conversion efficiency is of high concern. Switching regulators have a typical operating efficiency of 60–90%. In addition, they provide good regulation over a wide range of input voltage and maintain high efficiency over a wide range of load current. It is also possible to provide regulated voltages *larger* than the input voltage. A switching regulator has a significant size and weight advantage over other regulators in high-power applications: typically a switching regulator is smaller by a factor of between four and eight.

The basic topology of a switching regulator, as shown in Figure 14.15, is similar to a linear regulator except the active element is modulated in an ON/OFF mode of operation by a switch control unit. Typically, control unit uses the output of the error amplifier to pulse-width modulate the output of a constant-frequency oscillator. The pulse-width modulated square wave varies the ON/OFF duty cycle of the active element similarly varying the average value of the passed voltage. The addition of a rectifying low-pass filter after the series active element smoothes the output voltage to the desired value. Another realization of the switch control unit consists of a voltage-controlled oscillator with non-varying ON time. A major drawback of this realization is the added complexity of the rectifying filter due to variable frequency inputs. Consequently, the voltage-controlled oscillator realization is not typically found in modern switching voltage regulators.

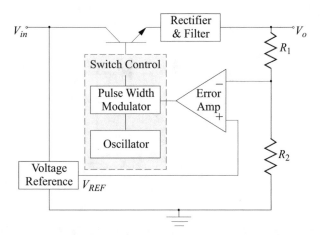

Figure 14.15: Typical switching regulator topology.

The added complexity of a switching regulator in the form of complex switch control circuitry and the need for a high-frequency active switching element increase the cost of the regulator. As a result, switching regulators can only compete economically with other regulators in high-power applications (greater than $\sim 20\,\text{W}$). Switching regulators typically have somewhat higher output ripple and can be more susceptible to load current transients than linear regulators. In addition, high rate (20–500 kHz), square-wave switching generates significant electromagnetic interference (EMI) and radio-frequency interference (RFI). It is possible successfully diminish both EMI and RFI with proper filtering.

The rectifying filtering section of a switching voltage regulator can take several different topologies. Most common among these topologies are:

- Buck

- Boost

- Buck-Boost

The basic topology of each rectifying filter section is shown in Figure 14.16. The majority of other rectifying filter are direct derivatives of these three types.

The output voltage, V_o, of the *buck* filter configuration (Figure 14.16a) is always less than the input voltage, V_{in}. Here, a switch is placed in series between the input voltage and the input to

---

[7]The active element switch is shown here as a BJT. In many switching regulator designs, is a much more complex element. Many designs use a class A amplifier with a transformer output as the high-frequency switch.

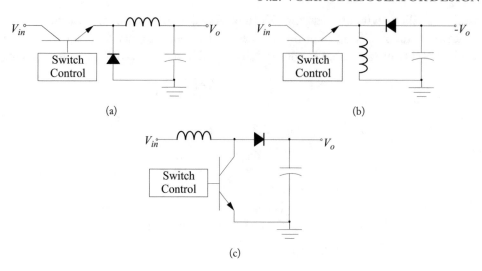

Figure 14.16: Switching regulator filtering configurations.[7](a) Buck (step down); (b) Buck-boost filtering (step up or down) (c) Boost filtering (step up).

an LC low-pass filter. When the switch is conducting, current flows through the inductor to the load. When the switch opens, the magnetic field within the inductor maintains the current flow to the load, pulling current through the diode from ground. In a buck circuit, the output voltage is proportional to the product of the input voltage and the switch duty cycle:

$$V_o \approx V_{in} \times \text{ (duty cycle)}. \tag{14.19}$$

Peak switch current in the buck circuit is proportional to the load current.

The *boost* filter configuration (Figure 14.16c) has the unique property of providing an output voltage that is always *larger* than the input voltage. In this circuit, the switch is placed within the LC filter so that it can shunt current to ground. With the switch in its conducting state, the inductor current increases. When the switch opens, the output voltage, V_o, is the sum of the input voltage and the voltage across the inductor (a positive voltage due to decreasing inductor current). Boost regulators deliver a fixed amount of power to the load:

$$P_L = \frac{1}{2}LI^2 f_o, \tag{14.20}$$

where I is the peak inductor current, and f_o is the switch operating frequency. In order to determine the output voltage the load resistance, R_L, must be known:

$$V_o = \sqrt{P_o R_L} = I\sqrt{\frac{L R_L f_o}{2}}. \tag{14.21}$$

Of course, the peak inductor current, I, is proportional to the duty cycle of the switch operation. Boost circuits are particularly useful in charging capacitive circuits (as in a capacitive-discharge automotive ignition system) and make good battery chargers.

The *buck-boost* rectifying filter (Figure 14.16b) provides the possibility of output voltages that are either higher or lower than the input voltage. The circuit operates in much the same fashion as the boost circuit with the exception that the output voltage is simply the voltage across the inductor. The buck-boost circuit also delivers constant power to the load, independent of the load resistance: Hence, Equations (14.20) and (14.21) are valid. The buck-boost has the distinct feature of providing a negative voltage output. This change of polarity is often an advantage, sometimes a drawback. Isolation of the input from the output through transformer coupling avoids any problems.

For all these configurations, transient changes in the load conditions may cause problems. If the load suddenly becomes a very high impedance or suddenly becomes disconnected, the energy stored in the inductor has no path for dissipation. In a worst case scenario, arcing across the load may occur. Switching power regulators are currently in a constant state of change due to major innovations in component design. It appears that major improvements in design are near.

14.3   VOLTAGE REGULATOR APPLICATIONS

The three-terminal, linear, IC voltage regulator is the most common voltage regulator in electronic applications with power requirements less that $\sim 20\,\mathrm{W}$ and voltages less than $50\,\mathrm{V}$. Its small size, high quality, and cost effective properties make it an extremely useful device. In its most simple form, the three-terminal regulator requires a minimum number of external components for proper operation: typical connection of a positive, fixed-output regulator is shown in Figure 14.17. The only external components necessary for this particular regulator are two small-value, high-frequency capacitors to improve stability and transient response.

Figure 14.17: Typical fixed IC voltage regulator connection.

Whenever using an IC voltage regulator, the circuit designer is faced with several design choices based on the properties and limitations of these regulators. Linear regulators are categorized by their regulated output voltage. The most basic categories are based on the following properties:

• Output Voltage Polarity

- Fixed or Variable Output Voltage

- Dual-tracking Output Voltages

The maximum output current is one additional constraint that must be considered. IC voltage regulators typically come with maximum current ranging from 100 mA to 3 A. It is possible to extend this maximum output current with the addition of external circuit pass elements (Section 14.3.1).

The *polarity* of the input and output voltage usually determines the use of a positive or negative regulator: positive regulators typically are used to regulate positive voltages—negative regulators typically regulate negative voltages. This is particularly true in systems where the input and output share a common ground. However, in systems where the ground reference can be floating at either the input or the output, the positive and negative regulators may be interchanged (Figure 14.18). In this special case a positive regulator can be used to regulate negative voltages and a negative regulator can be used to regulate positive voltages.

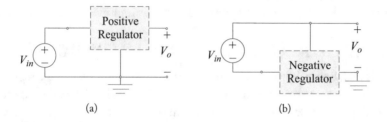

(a)   (b)

Figure 14.18: Voltage regulation alternatives. (a) Positive output using positive regulator; (b) Positive output using negative regulator.

Fixed output   voltage regulators are available in a variety of output voltage values and current ratings. They provide an inexpensive, simple means of regulating output voltage and have several advantages:

- ease of use

- few external components

- reliable performance

- internal thermal and short-circuit protection

The main disadvantage of a fixed output voltage regulator lies in the inability to precisely adjust its output. The variation in output voltage may be as large as ± 5% for any specified value. A similar problem exists due to the limited selection of output voltage values that are available.

Adjustable output voltage regulators are best suited for applications requiring high-precision voltage regulation and/or regulation at a non-standard voltage level. In addition, the regulated voltage may be sensed at a location remote from the output of the regulator. This feature allows for compensation due to losses in a distributed load or external pass components. Additional features often found on adjustable regulators include: adjustable short-circuit current limiting, access to the reference voltage, V_{REF}, and overload protection. Adjustable regulators typically require a few more external components than fixed regulators (Figure 14.19). The capacitors improve stability and transient response.

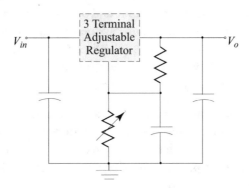

Figure 14.19: Typical adjustable voltage regulator connection.

It is possible to accomplish the performance of an adjustable voltage regulator using a fixed-output regulator and an OpAmp. One circuit topology that has an adjustable output is shown in Figure 14.20. In this circuit the variable output is limited to values larger than the specified,

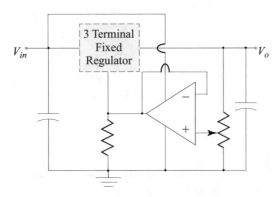

Figure 14.20: Variable regulated output using a fixed output voltage regulator.

regulated output of the fixed regulator. While designs of this type are effective, the additional components required often makes them economically impractical.

Many systems require balanced, dual-polarity power: an OpAmp that requires $\pm V_{CC}$ about a common ground is such a system. An obvious solution to dual-polarity applications is two independent regulators, one positive and one negative, paired together. Two problems arise with such a solution: power-ON latch-up and undervoltage output imbalance. Latch-up is due to the intolerance of each individual regulator to reverse voltages applied at its output. In dual-polarity systems with a single load across both outputs, reverse voltages may appear during the power-ON operation and cause latch-up of the output of one of the regulators. This condition can be avoided by placing diodes, from input to output and output to ground of each regulator, to avoid significant reverse voltage application. In many systems that require balanced power of opposite polarity, the application of unbalanced power will offset the output. If, for a variety of reasons, the imbalance is not constant, there will be a time-varying offset in the signal output: time-varying outputs are usually interpreted as information signals.

Dual Tracking   Regulators provide a solution to both problems. Latch-up is internally controlled and no additional external components are necessary. In order to avoid an imbalance in output voltages, the control system within a dual tracking regulator monitors both the positive and negative power outputs. If either output falls out of regulation, the tracking regulator will respond by varying the other output to match: a decrease in the magnitude of the positive output will result in an equal decrease in the magnitude of the negative output. A typical dual tracking connection is shown in Figure 14.21. As is the case with most linear regulators, output capacitors improve ripple and transient performance: input capacitors may be necessary if the source is particularly noisy or if the regulator is placed too far from the unregulated power supply.

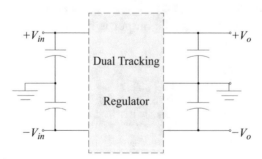

Figure 14.21: Typical dual tracking regulator connection.

Limitations as to the input voltage range and output voltage and current ranges over which regulation will occur apply to all voltage regulators. The *safe operating area* (SOA) defines the limits of these ranges. Exceeding the limits can result in catastrophic failure, temporary device shutdown, or failure to properly regulate the output. The SOA is defined by manufacturer's specifications relating to the input voltage, the output current, maximum power dissipation, and, in the case of variable voltage regulators, the output voltage. These specifications are described as:

- $V_{in(\text{max})}$ The absolute maximum input voltage with respect to the regulator ground terminal.
- $(V_{in} - V_o)_{\text{min}}$ The minimum input-output voltage difference at which regulation can be maintained. Also called the dropout voltage.
- $(V_{in} - V_o)_{\text{max}}$ The maximum input-output voltage difference.
- $I_{L(\text{max})}$ The maximum current deliverable to the load from the regulator.
- $P_{D(\text{max})}$ The maximum power that can be dissipated by the regulator.
- $V_{o(\text{min})}$ For adjustable regulators, the minimum output voltage that can be regulated.
- $V_{o(\text{max})}$ For adjustable regulators, the maximum output voltage that can be regulated.

Of these specifications, $V_{in(\text{max})}$, $I_{L(\text{max})}$, and $P_{D(\text{max})}$ can result in catastrophic failure if proper protection is not provided. Often this protection is within the regulator itself: in some cases it must be provided with external circuitry. The other specifications are functional limits that, if exceeded, imply a failure in the regulation ability of the device.

14.3.1 EXTENDING THE CAPABILITIES OF A SIMPLE VOLTAGE REGULATOR

It is often desirable to extend the SOA of a regulator with external components so as to exceed one of the specified limits. Extending the maximum values of output current, input voltage, and output voltage is necessary in many voltage regulator applications. While many, varied uses of simple IC voltage regulators exist, a constant current source provides a good example of one possibility.

Increasing Maximum Output Current

The effective output current of a simple voltage regulator can be increased through the use of an external bypass element as shown in Figure 14.22.

In this realization the bypass element is a power PNP BJT: it must be chosen with the capability to provide ample output current and to dissipate enough power across its collector-emitter terminals. The resistor, R, must be chosen so that the internal bias current of the voltage regulator, I_{Bias}, does not turn on the external bypass BJT. This condition is met if

$$R \leq \frac{V_{BE(on)}}{I_{Bias}}.$$

Special care must be taken in this realization to protect against a possible output short circuit. Under this condition, the bypass element often must dissipate large quantities of power and may fail.

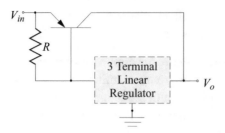

Figure 14.22: A series pass element to boost output current.

Increasing Input and/or Output Voltage Levels

It is often necessary to provide a regulated voltage larger than an available linear regulator. One possible circuit topology that will increase the regulated voltage for a particular regulator is shown in Figure 14.23. This circuit uses a zener diode in between the ground terminal of the regulator and the actual circuit ground. The new regulated output voltage is sum of the regulator output voltage and the series zener voltage:

$$V_{o(new)} = V_{o(reg)} + V_{z1}.$$

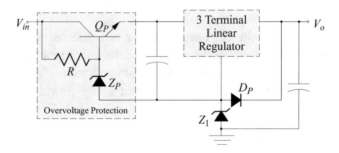

Figure 14.23: High output voltage output circuit topology.

Also shown in Figure 14.23 is an overvoltage protection circuit (shaded area and the diode D_P). Under normal operation the input BJT, Q_P, is in saturation and the circuit operates as if the protection circuit was not present. However, if the input voltage exceeds the sum of the two zener voltages,

$$V_{in} > V_{z1} + V_{zP},$$

the input BJT, Q_P, will enter the forward-active region begin to dissipate power. The voltage drop across the BJT collector-emitter terminals will protect the input of the regulator from overvoltage, and excess current will be shunted through R_p, Z_p and Z_1. The protection diode, D_p, protects the regulator against an output short circuit. While this protection circuit is shown in conjunction with the zener diode realization of extending the output voltage, it can be used alone. With the absence of zener diode, Z_1, the diode, D_P, is also excluded. All elements in this overvoltage protection circuit will dissipate large quantities of power when activated and must be rated for that occurrence.

Using a Voltage Regulator to Provide a Constant Current

Another common regulated source necessary in many electronic applications is a constant current source. Three terminal positive voltage regulators can be effectively used to provide this source as shown in Figure 14.24.

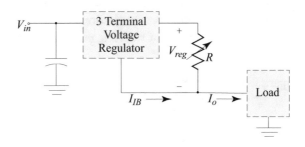

Figure 14.24: Adjustable constant current source regulator configuration.

In this configuration the output current, I_o, can be adjusted to any value from the minimum regulator bias current (≈ 8 mA) to the maximum current deliverable be the regulator, $I_{L(\max)}$. The output current is the sum of current through the variable resistor and the bias current through the ground terminal of the regulator:

$$I_o = \frac{V_{reg}}{R} + I_{IB}.$$

The input voltage for this configuration must always be greater than the sum of the minimum input-output voltage for the regulator, the regulator voltage, and the voltage at the load:

$$V_{in} \geq (V_{in} - V_o)_{\min} + V_{reg} + V_o.$$

The choice of regulator is a balancing of conflicting requirements. Small V_{reg} is desirable to reduce power dissipation and allow for large variation in V_o. Large V_{reg} gives a more precise setting of the load current through the variable resistor. Five volt fixed regulators often provide a good compromise.

14.4 TRANSIENT SUPPRESSION AND OVERVOLTAGE PROTECTION

A common problem in most electronic applications comes in the form of an inappropriately large input voltage. Protection must be provided so that erroneous voltage application connection at the input or transient overvoltages do not damage the circuitry. Transient high voltages are usually the result of the release of stored electric energy in the form of a current or voltage pulse. In order to ensure proper circuit protection against improper operation or actual damage, the energy within the transient must be dissipated within an added suppressor. Transients may result from sources within a circuit or have external sources. One typical internal source is inductive switching (as in motor control circuitry). External sources of transients include line-voltage variations and electromagnetic interference (EMI). One particularly unfriendly environment (in terms of transients) for electronic applications occurs in automotive applications.

Common Transient Suppressors and/or Overvoltage Protection Circuits limit the peak AC or DC voltage and include:

- Carbon Block Spark Gap Suppressors

- Zener Diodes

- Varistors

- Selenium Rectifiers

- Overvoltage "Crowbar" Circuits

The first four items on this list are well suited as protection against short duration high-voltage transients. In addition, zener diodes and Selenium rectifiers have good capability to protect against longer duration overvoltages so long as their average power ratings are not exceeded. Overvoltage crowbar circuits are best used as protection against long duration overvoltages or misapplication of electrical power. Each of these protection circuits is connected as a shunt to the circuit to be protected.

A *carbon block spark gap suppressor* is an effective means of keeping transient voltages below a level of 300–1000 volts. It is commonly found in telecommunication and power distribution system. The suppressor is simply two carbon electrodes, one connected through a resistor to ground, separated by an air gap of approximately 0.1 mm. When a high voltage transient is encountered, the energy in the transient is passed to ground through the arc in the air gap between the electrodes. The main drawbacks to carbon block spark gap suppresser use is its relatively short lifetime and the inherent variation in arcing voltage.

Zener diodes, discussed in Chapter 2 (Book 1), exhibit excellent voltage limiting properties. They are perhaps the most often used transient suppressor. Often a "back-to-back" connection, as shown in Figure 14.25, is used to suppress transient overvoltages that are bipolar in nature: many zener transient voltage suppressors are available commercially packaged as a back-to-back

pair. While typical zener diodes are designed to operate at less than their maximum power rating, zener transient suppressors are designed to effectively limit large, short-duration power pulses. The design change is accomplished by increasing junction area so as to withstand the high energy surge of a transient. When zener diodes fail, they typically fail to a short circuit. This feature, coupled with the shunt connection of zener protection circuits, ensures that the circuit is protected even though the protection circuit has failed.

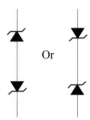

Figure 14.25: Zener diodes connected for bipolar overvoltage protection.

A *varistor* is a non-linear resistor that has electrical properties similar to a back-to-back zener diode connection. A varistor can therefore provide an alternative to zener diodes in many applications. The device is constructed of metal (often zinc) oxides sintered into a polycrystalline structure or silicon carbide sintered into a suitable ceramic binder forming a hard ceramic-like material. As a result of the sintering process, highly resistive intergranular boundaries are distributed throughout the volume of the device. The boundaries are the source of the voltage-dependent non-linear behavior of the varistor. The distributed nature of the boundaries make the varistor well suited for high power transient suppression. Compared to zener diodes, varistors are inexpensive. Varistors also share the attractive fail-to-short characteristic with zener diodes.

Figure 14.26: Varistor circuit symbol.

The non-linear behavior of a varistor can be described by its volt-ampere transfer relationship:

$$V = \frac{C}{I} |I|^{(2-a)}, \qquad (14.22)$$

where the passive sign convention relates the voltage and current polarities[8] and the two varistor property-dependent constants are given by:

$$C = \text{the voltage across the varistor at } I = 1\,\text{A}$$
$$a = \text{a factor describing the non-linearity of the V-I curve. Typically,}$$
$$0.5 \leq a \leq 0.83 : \text{ as } a \text{ increases, the sharpness of the curve increases.}$$

In Figure 14.27, the volt-ampere transfer relationship of a typical varistor ($a = 0.7$) is compared to that of a back-to-back zener diode connection. The difference in the sharpness of the cutoff is evident.

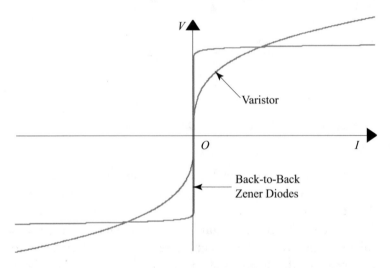

Figure 14.27: Volt-ampere relationship for a varistor compared to a back-to-back Zener diode pair.

The major drawbacks to varistors can be significant. Varistors do not clamp voltages as effectively as zener diodes and may, therefore, be unsuitable in many applications where an absolute maximum voltage limit must be maintained. Also, they can only dissipate relatively small amounts of average power and are degrade significantly when stressed near their maximum ratings. They can, however, offer a significant cost advantage over zener diodes. Varistors are best used in applications for protection against rarely occurring transients.

Selenium rectifiers, in the reverse breakdown mode of operation, can clamp voltages in much the same manner as zener diodes. An attractive feature of many Selenium rectifiers is the ability to survive a limited number of surges greater than their maximum typical rating. Unfortunately they do not have as sharp a "knee" in the V-I curve as the zener diode (though sharper than the varistor)

---

[8]The volt-ampere transfer relationship for a varistor is symmetric. The literature typically provides a relationship that assumes current flow is always positive: $V = CI^{(1-a)}$.

and the ON resistance is somewhat greater. Therefore, the use of Selenium rectifiers as overvoltage protection devices is diminishing as the demands for protection become more stringent.

Overvoltage "Crowbar" Circuits   protect loads with a switchable shunt element. This element is often realized with SCR that is activated if overvoltage conditions exist. A simple form of overvoltage crowbar circuit, utilizing a zener diode as the overvoltage sensing element, is shown in Figure 14.28. In this circuit, an input voltage, V_{in}, that exceeds the SCR gate activation voltage plus the zener breakdown voltage, will force current into the SCR gate activating the SCR. The SCR will then shunt current away from the load protecting it from damage. One drawback of SCR crowbar circuits is that they will not deactivate unless the source current goes to zero. Deactivation of the SCR is usually accomplished by a series circuit breaker or fuse incorporated in series with the source, V_{in}.

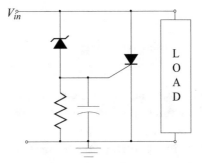

Figure 14.28: A simple SCR crowbar circuit.

While a zener diode—SCR crowbar circuit is relatively inexpensive and easy to use, there are several drawbacks to its use. These drawbacks are mainly due to the properties of zener diodes. In particular, the zener voltage values commercially available are limited, often have insufficient tolerances, and may not activate sharply enough (the knee of the diode curve may be too rounded). These drawbacks are particularly significant when the voltage protection limit must be fairly small: many digital circuits require overvoltage protection at power voltages less than 10 V. A good solution to these problems involves the use of an integrated circuit sensing circuit.

An overvoltage crowbar circuit using an IC sensing circuit is shown in Figure 14.29: the shaded portion of the figure highlights the sensing circuit. In this IC package are contained two OpAmps, a voltage reference circuit, a zener diode, a current source, and three BJTs. The operation of the circuit is reliable and accurate.

In the normal, deactivated state, the input voltage to the protection circuit, V_{in}, is appropriately small:

$$V_{in} < \frac{R_2 + R_1}{R_2} V_{REF}.$$

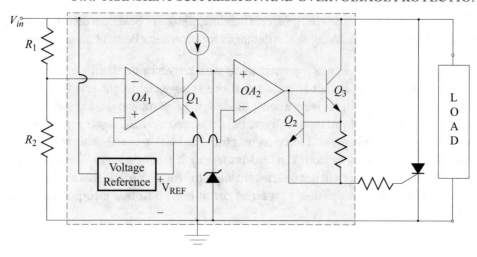

Figure 14.29: Crowbar protection using overvoltage sensing circuit.

This deactivated state assures that the first OpAmp, OA_1, provides a high voltage to the base of Q_1, putting it into the saturation region. OA_2 then provides a low voltage to Q_2 and Q_3 which are in cutoff. No current is supplied to the gate of the SCR which is, therefore, not activated. Should V_{in} increase so that the input voltage fails to meet the constraints of Equation (14.22), OA_1 will turn off Q_1. The current source will then pass its current through the zener diode raising the positive terminal of OA_2 higher than V_{REF}. The BJT combination Q_2 and Q_3 will turn on and activate the SCR, shunting all the current away from the load.

Using an IC overvoltage sensing circuit to activate the SCR allows the circuit designer to provide a temperature independent voltage reference, and adjust the crowbar voltage using the two resistors, R_1 and R_2. The sensing circuits are readily available from a wide range of manufacturers and are usually found within the "power supply supervisory" listings. It is also common to package a circuit equivalent to that shown in Figure 14.29 in a single IC package. These crowbars are available in a range of voltages and short-circuit current ratings.

While it is impossible to demonstrate the operation of every protection circuit, the protection methods discussed here provide a good sampling of typical techniques. Zener diodes provide appropriate transient protection in most small-scale electronic applications. Higher power applications that need protection from longer duration overvoltages usually use crowbar devices. Crowbar circuits are also often activated by temperature sensors (often a thermistor) to protect against damage to a circuit due to overheating. IC overvoltage sensors occasionally are used to deactivate a semiconductor switch in series with the load rather than the shunt SCR as described here.

14.5 CONCLUDING REMARKS

The study of power electronics should, more properly, be covered in an entire book rather than in one short chapter. Here, the discussion has been restricted to a few common devices that an electronic circuit designer will find necessary in order have clean, constant DC power as an input to circuits with other major electronic functions. Thyristors were shown to be a useful family of triggered switches dominated by the SCR and Triac. Both linear and switched voltage regulators are commonly used: linear regulators dominate low-power applications, while switched regulators find greater use in high-power applications. Several forms of active and passive protection circuits were also discussed.

Summary Design Example

In the design of Analog-to-Digital (A/D) and Digital-to-Analog (D/A) converters. there is a need for a precision voltage reference that is relatively invariant with input voltage and temperature variation.

Design a 2.5 V precision voltage reference that will operate for input voltages ranging from 4 to 10 V.

Solution:

The three voltage reference topologies available are:

- Zener diode

- Constant-current Zener diode

- Bandgap

The basic Zener diode reference is somewhat dependent on input voltage and must, therefore, be discarded. The constant-current Zener diode reference is a possibility, but the relatively small possible difference between input an output voltage may pose a problem. Therefore, the bandgap reference seems the best choice and will be form the basis of the chosen design.

As described in Section 14.2, the bandgap reference output voltage is given by:

$$V_{REF} \approx \eta \, V_t \frac{R_3}{R_2} \ln \left(\frac{I_{C1}}{I_{C2}} \right) + V_{BE3}.$$

Assuming that $V_{BE3} = V_\gamma = 0.7 \, \text{V}$, this expression reduces to:

$$2.5 = 0.026 \frac{R_3}{R_2} \ln \left(\frac{I_{C1}}{I_{C2}} \right) + 0.7,$$

or

$$\frac{R_3}{R_2} \ln \left(\frac{I_{C1}}{I_{C2}} \right) = 69.231.$$

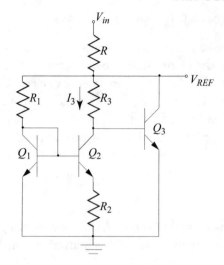

A few arbitrary design choices must be made to continue. Choose

$$I_{C2} = 0.5 \, \text{mA} \qquad \Rightarrow \qquad R_3 = \frac{2.5 - 0.7}{I_{C2}} = 3.6 \, \text{k}\Omega.$$

Also choose

$$I_{C1} = 1.0 \, \text{mA} \qquad \Rightarrow \qquad R_2 = 36.0 \, \Omega,$$

and

$$\Rightarrow \qquad R_1 = \frac{2.5 - 0.7}{I_{C1}} = 1.8 \, \text{k}\Omega.$$

The resistor, R, must be small enough so that at least 1.5 mA ($I_{C1} + I_{C2}$) will flow at the smallest input voltage. The upper limit on input voltage merely puts a restriction on the power dissipation. Therefore,

$$R < \frac{4 \, \text{V} - 2.5 \, \text{V}}{1.5 \, \text{mA}} = 1 \, \text{k}\Omega \quad \Rightarrow \qquad \text{Choose } R = 910 \, \Omega.$$

The maximum power dissipated in the resistor R is:

$$P_{R(\text{max})} = \frac{(10 - 2.5)^2}{910} = 61.8 \, \text{mW}.$$

A 1/4 W resistor will suffice.

Any high-β_F BJTs can be used for this design. The ideal temperature variation for the BJTs is given by the relationship of Equation (14.16). This expression reduces to:

$$\frac{\Delta V_{BE3}}{\Delta T} = -\frac{11600}{\eta} \left\{ \frac{R_3}{R_2} \ln \left(\frac{I_{C1}}{I_{C2}} \right) \right\} = -5.98 \, \text{mV/°K}.$$

This is a rather large value: more realistic BJTs will provide a voltage with small, but non-zero temperature variation.

14.6   PROBLEMS

14.1.   Model the action of an SCR using the equivalent circuit of Figure 14.3b and SPICE. Use BJTs with $\beta_F = 100$. Apply a voltage, V_s, of 18 V_{p-p} at 60 Hz in series with 100 Ω across the anode-cathode terminals. Inject into the gate of the SCR model, 10 μA current pulses of 50 μs duration from a source with an output resistance of 100 kΩ. Verify the operation of the SCR model as shown in Figure 14.3.

14.2.   A macromodel for the 2N1599 SCR is available in the SPICE model libraries. Its terminals are ordered in the model call as {anode gate cathode}. Repeat the functional test of Problem 14.1 (the minimum gate trigger current is 2 mA: increase the magnitude of the current pulses so that the SCR will properly trigger).

14.3.   A macromodel for the 2N6073 Triac is available in the SPICE model libraries. Its terminals are ordered in the model call as {MT2 gate MT1}. Repeat the functional test of Problem 14.1 on a 100 V_{p-p} sinusoid to show that *positive* gate current pulses trigger conduction in both directions (the minimum magnitude gate trigger current is about 20 mA: increase the magnitude of the current pulses so that the SCR will properly trigger).

14.4.   A macromodel for the 2N6073 Triac is available in the SPICE model libraries. Its terminals are ordered in the model call as {MT2 gate MT1}. Repeat the functional test of Problem 14.1 on a 100 V_{p-p} sinusoid to show that *negative* gate current pulses trigger conduction in both directions (the minimum gate magnitude trigger current is about 20 mA: increase the magnitude of the current pulses so that the SCR will properly trigger).

14.5.   A power conversion system is under design. The input to this system is standard 110 V_{AC} at 60 Hz: the output is variable-voltage DC supplied to a 100 Ω resistive load. It has been decided to achieve the design goals using a variable delay one-shot and an SCR as shown.

   (a) Complete the design using a 2N1599 SCR and a capacitor that will provide no more than 8% ripple.

   (b) Use SPICE to verify proper operation when the SCR is triggered over 20% of the input sinusoidal waveform.

   (c) Use SPICE to verify proper operation when the SCR is triggered over 80% of the input sinusoidal waveform.

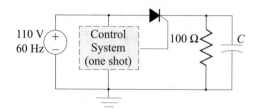

14.6. The power conversion system of Problem 14.5 is to be redesigned using a Triac rather than an SCR. This redesign requires the addition of a full-wave rectifier bridge, but allows greater efficiency of power conversion.

(a) Complete the design using a 2N6073 Triac and a capacitor that will provide no more than 8% ripple.

(b) Use SPICE to verify proper operation when the Triac is triggered over 20% of the input sinusoidal waveform.

(c) Use SPICE to verify proper operation when the Triac is triggered over 160% of the input sinusoidal waveform.

14.7. The basic Zener diode voltage reference circuit of Figure 14.10 is proposed as a voltage reference circuit with an output of 4 V. Assume the Zener diodes are characterized by a Zener voltage, $V_z = 4$ V, and Zener resistance. $r_z = 40\,\Omega$. The resistor, R, in the voltage reference circuit has value $10\,\text{k}\Omega$.

(a) Determine the nominal output voltage, V_{ref}, of the circuit to an input, $V_{in} = 16$ V.

(b) Assume that V_{in} experiences an AC ripple of amplitude $2\,V_{p-p}$. What is the AC ripple in the output, V_{ref}?

14.8. The circuit shown is proposed as a voltage reference circuit with an output of 4 V. Assume the Zener diodes are characterized by a Zener voltage, $V_z = 4$ V, and Zener resistance. $r_z = 40\,\Omega$.

(a) Determine the nominal output voltage, V_{ref}, of the circuit to an input, $V_{in} = 16$ V.

(b) Assume that V_{in} experiences an AC ripple of amplitude $2\,V_{p-p}$. What is the AC ripple in the output, V_{ref}?

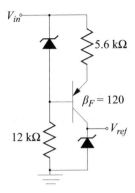

14.9. Design a constant-current voltage reference with an output of 2.0 V. Assume an input voltage ranging from 4 V to 6 V, Silicon BJTs and Zener diodes with the following properties:

minimum Zener current	—	1 mA
Zener resistance	—	5 Ω
β_F	—	150

14.10. A constant-current voltage reference is designed with the basic topology identified in Figure 14.11 and the following circuit elements (parameters are found in the SPICE libraries):

BJTs	— 2N2222	Zener diode	— 1N750
R_{SC}	— 680 Ω	R	— 1.6 kΩ

(a) Estimate the nominal output voltage.

(b) Assume the input voltage, V_{in}, varies between 8 V and 12 V, use SPICE to determine the variation in the output voltage, V_{ref}.

(c) Compare SPICE results to theory.

14.11. Design a bandgap voltage reference with an output of 1.2 V. Assume an input voltage ≈ 5 V. What is the ideal temperature variation of the transistor base-emitter junction voltage?

14.12. Design a bandgap voltage reference with an output of 1.4 V. Assume an input voltage ≈ 12 V. What is the ideal temperature variation of the transistor base-emitter junction voltage?

14.13. Design a 5 V series voltage regulator using the basic topology shown in Figure 14.13. The input voltage falls in the range 6 V < V_{in} < 10 V. Assume the following components are available:

Precision voltage reference	— 1.2 V
Comparator	— LM111
Bipolar Junction Transistor	— $\beta_F = 100$
Resistors	— any standard value

Verify correct operation using SPICE for a load of 100 Ω. Note: the macromodel for the LM111 comparator has an open-collector output—a pull-up resistor is required for HIGH output.

14.14. Design a 3.3 V series voltage regulator using the basic topology shown in Figure 14.13. The input voltage falls in the range 6 V < V_{in} < 12 V. Assume the following components are available:

Precision voltage reference	— 1.1 V
Comparator	— LM111
Bipolar Junction Transistor	— $\beta_F = 120$
Resistors	— any standard value

Verify correct operation using SPICE for a loads of 100 Ω and 1 kΩ. Note: the macromodel for the LM111 comparator has an open-collector output—a pull-up resistor is required for HIGH output.

14.15. Design a 5 V shunt voltage regulator using the basic topology shown in Figure 14.14. The input voltage falls in the range 6 V < V_{in} < 10 V. Assume the following components are available:

Precision voltage reference	— 1.2 V
Comparator	— LM111
Bipolar Junction Transistor	— $\beta_F = 100$
Resistors	— any standard value

Verify correct operation using SPICE for a load of 100 Ω. Note: the macromodel for the LM111 comparator has an open-collector output—a pull-up resistor is required for HIGH output.

14.16. Design a 3.3 V shunt voltage regulator using the basic topology shown in Figure 14.14. The input voltage falls in the range 6 V < V_{in} < 10 V. Assume the following components are available:

Precision voltage reference	— 1.2 V
Comparator	— LM111
Bipolar Junction Transistor	— $\beta_F = 100$
Resistors	— any standard value

Verify correct operation using SPICE for a load of 100 Ω. Note: the macromodel for the LM111 comparator has an open-collector output—a pull-up resistor is required for HIGH output.

14.17. Use SPICE and the macromodel for the LM7805C to verify that a + 5 V regulator can correctly regulate to achieve a + 5 V output.

   (a) Over what range of output current will the LM7805C macromodel provide correct regulation?

   (b) What is the minimum input voltage that provides correct regulation for an output current of 100 mA? 10 mA?

14.18. Use SPICE and the macromodel for the LM7815C to verify that a + 15 V regulator can correctly regulate to achieve a + 15 V output.

   (a) Over what range of output current will the LM7815C macromodel provide correct regulation?

   (b) What is the minimum input voltage that provides correct regulation for an output current of 100 mA? 10 mA?

14.19. Use SPICE and the macromodel for the LM7805C to verify that a + 5 V regulator can correctly regulate to achieve a − 5 V output. Assume the input voltage varies in the range, $-7 < V_i < -15$ and a load current of 150 mA.

14.20. Design a 7.5 V voltage regulator using a 5 V, three-terminal regulator and the basic topology shown in Figure 14.20. The input voltage falls in the range 9 V < V_{in} < 15 V. Assume the following components are available:

5 V fixed voltage regulator	— LM7805C
OpAmp	— μA741
Resistors	— any standard value

Verify correct operation using SPICE for a load of 100 Ω.

14.21. Design a 16 V voltage regulator using a 12 V, three-terminal regulator and the basic topology shown in Figure 14.20. The input voltage falls in the range 20 V < V_{in} < 30 V. Assume the following components are available:

12 V fixed voltage regulator	— LM7812CT
OpAmp	— μA741
Resistors	— any standard value

Verify correct operation using SPICE for a load of 200 mA.

14.22. An automotive application requires a current of $10 \pm 0.1\,\text{mA}$ to be delivered to a variable load ($470\,\Omega > R_{load} > 100\,\Omega$). The input voltage to the system varies between 11 V and 16 V. Design a constant current source using a LM7805C 5 V regulator. Verify proper operation of the design using SPICE.

14.23. The 78XX series of voltage regulators is limited to approximately 1.0 A output current (with appropriate heat sinks). The current boost configuration of Figure 14.22 has been selected for a design requiring a regulated output voltage at an output current, $I_o \approx 5.0\,\text{A}$. The input voltage is: $V_{in} \approx 10\,\text{V}$. Complete the design by specifying proper resistance values and component power ratings. Assume $\beta_F = 100$.

14.24. The 78XX series of voltage regulators is limited to approximately 1.0 A output current (with appropriate heat sinks). The current boost configuration shown has been selected for a design requiring a regulated output voltage at an output current, $I_o \approx 5.0\,\text{A}$. The input voltage is: $V_{in} \approx 10\,\text{V}$. Complete the design by specifying proper resistance values and component power ratings. Assume $\beta_F = 100$.

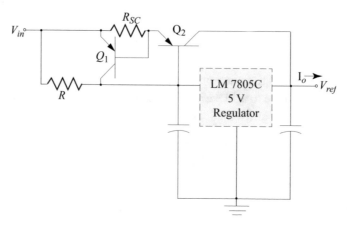

14.25. Complete the design of the overvoltage protection circuit shown by specifying the power ratings necessary for Q_p and Z_p if the input voltage is limited to 50 V. Use SPICE to verify proper operation. Assume $\beta_F = 60$ and a maximum load current of 250 mA.

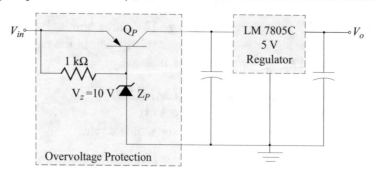

14.26. It is necessary to provide a regulated voltage at 10 V with a maximum current of 100 mA. The input voltage lies in the range, $15 < V_i < 55$ V. A LM7805, 5 V regulator is the only readily available component. While this regulator can supply adequate current, the improper output voltage and an input voltage limitation of 35 V demands that additional circuitry be added to the design.

   (a) Using the basic circuit topology of Figure 14.23, design a voltage regulator that will meet specifications. Be sure all component specifications include power ratings. Assume $\beta_F = 60$.

   (b) Verify proper operation of the design using SPICE.

14.27. A Varistor that has a non-linearity factor, $a = 0.80$, and allows one ampere or current at 50 V is being compared to back-to-back Zener diodes with SPICE parameters, IS = 10 nA, IBV = 50 mA, and BV = 15 V.

   (a) At what voltage does each voltage protection system allow the same current?

   (b) What is the value of that current?

   (c) If the current is increased 20% what voltage appears across each system?

14.28. Back-to-back 1N750 Zener diodes are proposed as a transient protection device for a 5 V circuit. The device being protected is essentially resistive and draws a nominal current of 50 mA. Use SPICE to determine the following.

   (a) The maximum voltage that will be applied to the load if an input current twice the nominal value is applied to the protected device.

   (b) The maximum voltage that will be applied to the load if a current spike of magnitude 150 mA and duration 0.1 μs (rise and fall times $\approx$ 0.1 ns) is applied to the load.

14.29. An approximate model of a varistor can be generated in SPICE using a non-linear voltage-controlled current source. The controlling voltage for the source is simply the voltage across the source. The non-linear properties of the varistor can be approximated as an odd-order polynomial in the voltage across the varistor.

   (a) For a varistor that has a non-linearity factor, $a = 0.80$, which allows one ampere of current at 50 V, determine an approximate expression for the current through the varistor expressed as a fifth-order polynomial of the voltage across the varistor.

   (b) Use SPICE to plot the V-I transfer relationship of this model.

   (c) Compare this V-I transfer relationship to that of back-to-back Zener diodes with SPICE parameters IS = 10 nA, BV = 15 V, and IBV = 50 mA.

   (d) Compare the voltage across each system when subjected to a current consisting of a 2 mA constant current that momentarily (duration 1 μs) jumps to 4 mA.

14.30. Design a simple SCR crowbar overvoltage protection circuit that activates at an input voltage of + 10 V. Use the following parts:

Zener diode — any specified Zener voltage
Resistors — any standard value
SCR — 2n1599 (gate turn-on voltage ≈ 0.7 V: turn-on current ≈ 2 mA)

Assume a source output resistance of 10 Ω.

Verify correct crowbar operation using SPICE for a load of 100 Ω. Comment on the accuracy of the crowbar voltage and what variation in design might be necessary.

14.31. Design a simple SCR crowbar overvoltage protection circuit that activates at an input voltage of + 40 V. Use the following parts:

Zener diode — any specified Zener voltage
Resistors — any standard value
SCR — 2N1599 (gate turn-on voltage ≈ 0.7 V: turn-on current ≈ 2 mA)

Assume a source output resistance of 10 Ω.

Verify correct crowbar operation using SPICE for a load of 1 kΩ. Comment on the accuracy of the crowbar voltage and what variation in design might be necessary.

REFERENCES

[1] ——-, *Linear Circuits Data Book, Volume 3, Voltage Regulators and Supervisors*, Texas Instruments Inc., Dallas, 1989.

[2] ——-, *Linear/Switchmode Voltage Regulator Handbook, 4th Ed.*, Motorola Inc., Phoenix 1989.

[3] ——-, *Thyristor Device Data Manual*, Motorola, Inc., Phoenix, 1992.

[4] Fisher, Marvin J., *Power Electronics*, PWS-Kent Publishing Co., Boston, 1991.

[5] Baliga, Jayant B. and Chen, Dan Y., Editors, *Power Transistors: Device Design and Applications*, the Institue of Electrical and Electronics Engineers, Inc., New York, 1984.

[6] Cherniak, Steve, *A Review of Transients and Their Means of Suppression*, Motorola, Inc., Phoenix 1991.

[7] Horowitz, Paul and Hill, Winfield, *The Art of Electronics, 2nd Ed.*, Cambridge University Press, Cambridge, 1989.

[8] Kassakian, John G., Schlecht, Martin F., and Verghese, George C., *Principles of Power Electronics*, Addison-Wesley Publishing Co., Reading, 1991.

[9] Millman, Jacob, *Microelectronics, Digital and Analog Circuits and Systems*, McGraw-Hill Book Co., New York, 1979.

[10] Mitchell, Daniel M., *DC - DC Switching Regulator Analysis*, McGraw-Hill Book Company, New York, 1988.

CHAPTER 15

Communication Circuits

The rapid development of computer technology has been followed by an explosion in the demand for high-speed low-cost telecommunication products to complement the growing demand for information transfer. New communication systems have been exploited to satisfy the telecommunication needs of the public. For example, the rapid growth in private wireless communication, most notably the cellular telephone, has increased the demand for high quality, low cost electronic circuits.

As this chapter will show, many subsystems that make up modern communication systems are designed using the basic principles established in the previous chapters. By integrating functional electronic blocks consisting of simple amplifiers or filters, application-specific communication electronic subsystems can be designed. Although a complete discussion of communication electronics requires volumes of books, this chapter presents an overview of some of the concepts used in designing those circuits.

The discussion in this chapter is limited to circuits operating at modest frequencies (below 20 MHz). In the *RF* range of frequencies, a different set of two port parameters, called *S-parameters*, are used to quantify active and passive devices and circuits. *S*-parameters are based on the devices or circuit behavior when terminating the network ports with a known impedance rather than an open circuit or a short circuit. Another reason for using *S*-parameters lies in the dependence of *RF* and microwave circuit analysis on the reflection (or Scattering, hence *S*-parameter) of electromagnetic waves.

The topics in this chapter are representative of the types of circuits commonly used in communications systems. Overviews are presented on analog-to-digital conversion, voltage-controlled oscillators, mixers, phase-lock loops, filter concepts, modulator/demodulator design, and issues important to good receiver design.

15.1 ANALOG-TO-DIGITAL CONVERSION

Most of the chapters in this text have been concerned exclusively with digital or exclusively analog topics. However, many systems use mixed-mode signaling: mixed-mode systems process both digital and analog signal characteristics. Consequently, the conversion of analog-to-digital signals is a common practice in modern electronic systems.

For example, audio signals in the form of music are converted to digital signals using sophisticated analog-to-digital converters (ADC) and stored on compact disks (CD) for sale to the public. Analog telephone voice signals are converted to digital signals for transmission on fiber optic systems. High Definition Television (HDTV) signals used digital signaling to transmit high resolution video images.

As technology advances, more signals from various transducers (microphones, cameras, medical sensors, and other sensors) are converted from analog to digital form for efficient and reliable transmission.

There are many different ADC circuits. Many incorporate digital integrated circuits which have not been discussed in this book. Therefore, the discussion in this chapter is limited to the Flash ADC circuit. However, prior to describing the Flash ADC, an introduction to analog-to-digital conversion principles is in order.

15.1.1 QUANTIZING AND SAMPLING

An analog signal voltage range is defined as the maximum peak-to-peak voltage excursion of the signal as shown in Figure 15.1a. When the peak-to-peak amplitude of the analog signal is divided into 2^N equal quantization levels (or bands) and sampled at times $T_0, T_1, T_2, \ldots, T_n$, discrete values of the signal voltage at the sampled times can be found. The result is a digitized form of the analog signal. The analog signal with its associated quantization levels and sampling times is shown in Figure 15.1b.

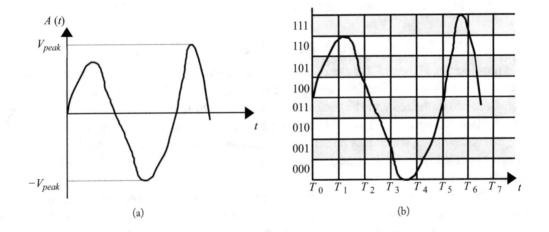

(a)   (b)

Figure 15.1: (a) Analog signal in time domain; (b) Quantizing times and sampling levels.

The resulting digital reconstruction of the analog signal in Figure 15.1 using $N = 3$ quantization levels at sampling times $T_0, T_1, T_2, \ldots, T_n$, is shown in Figure 15.2.

When the analog signal is sampled, the voltage level corresponding to the sampling time is held at that quantization level until the next sample causes the level to shift. Notice that each sampled quantization level takes a discrete value corresponding to the discrete quantizing levels (in Figure 15.2 there are 8 levels). Upon completion of the sampling process, the ADC generates

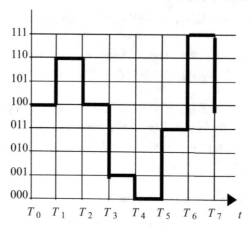

Figure 15.2: Digital signal reconstruction of analog signal.

a binary code representing the sampled quantization levels. In Figure 15.2, the number of discrete quantization levels is 2^N, where $N = 3$, so the ADC generates a binary code of 3 bits/sample. Each discrete quantization level in the case where $N = 3$ has amplitude $2V_{peak}/8$).

15.1.2 QUANTIZING ERROR AND SAMPLING FREQUENCY

The waveform in Figure 15.2 is clearly not an exact duplicate of the analog waveform shown in Figure 15.1. By low-pass filtering the digitized waveform, the high frequency components of the signal can be eliminated and the square corners rounded to improve its likeness to the original signal. However, because the sampled signal is forced to take on a discrete, quantized value, the original voltage level, which may lie between the discrete levels, may not be reproduced. This type of error is called quantizing error. In communications systems, this type of distortion is called quantizing error.

The quantizing error can be reduced by generating more discrete voltage levels and reducing the voltage sampling interval. For example, a 4-bit ADC converts an analog signal to 16 discrete quantization levels: 14-bit ADCs converts an analog signal to 16,384 discrete quantization levels. Although advantageous in many respects, high resolution (large number of quantization intervals, 2^N) creates a significant design problem when sampling small input signals due to random thermal noise.

Another common form of distortion occurs when the incoming analog waveform is sampled at too low a frequency, corresponding to an unacceptably large sampling interval. This problem is corrected by sampling at frequencies equal to or greater than required by the Nyquist sampling criterion. The Nyquist criterion states that if the highest frequency component of a waveform is f_h, then the waveform must be sampled at least at the Nyquist frequency, $2f_h$.

15.1.3 FLASH ANALOG-TO-DIGITAL CONVERTERS

Common ADC topologies used in modern communications systems include the single-ramp ADC, dual-ramp ADC, charge-redistribution ADC, and Flash ADC. All but the Flash ADC require digital logic implementations. The basic strategy of the single-ramp ADC is to convert the analog input voltage to a measurable time interval and transform that time interval to a digital word. The dual-ramp ADC is a higher resolution form of the single-ramp ADC. The operation of the charge-redistribution ADC is based on the cycling of charge about a binary weighted array of capacitors.

One of the simplest ADC circuits involves a parallel set of voltage comparators used to compare the incoming signal waveform voltage to a reference voltage. This form of ADC circuit is commonly called a *Flash ADC* and is shown in Figure 15.3. The input waveform has been processed by a sample-and-hold circuit like the one shown in Figure 15.4 and the output is typically decoded to become binary.

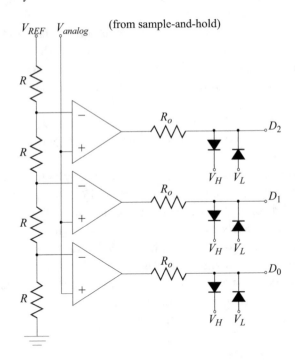

Figure 15.3: 2-Bit (4 level) flash ADC with clamps to limit output voltage.

The function of the sample-and-hold circuit is to track the input and hold the input voltage level providing a stable output signal to the ADC during the sampling interval. The basic storage device that "holds" the voltage in Figure 15.4 is the capacitor C. The capacitor charges rapidly when the switch is closed. When the switch, usually a FET controlled by digital circuitry,

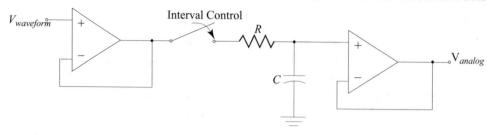

Figure 15.4: Sample-and-hold circuit.

is open, the large time constant of the capacitor and the input resistance of the output buffer OpAmp prevent the capacitor from discharging. Thus, a stable voltage is provided during the actual conversion process. For the circuit to operate properly, the charging time must be very short with respect to the input waveform variations. The charging and discharging times associated with the RC circuit are chosen to achieve the desired design accuracy and sampling time interval.

The operation of the parallel bank of comparators in a Flash ADC is based on the processing of the waveform in relation to the known reference voltage derived from precision voltage reference.[1] This reference voltage is divided by a series of resistors of equal value into the desired number of quantization intervals. Using the parallel bank of comparators, the input waveform sample is compared to each divided voltage: when the sample is larger, the comparator will output a positive result. All comparators with divided reference voltage below the input sample voltage will have positive outputs.

The high parallelism of the Flash ADC has the potential for very high-speed analog-to-digital conversion. However, the difficulties presented by this type of ADC include the large number of resistors and comparators required for a large number of quantization levels. Additionally, the output must be encoded into a binary code which requires additional circuitry. These encoders can be complex and may offset the high speed advantages of the Flash ADC.

Despite the potential drawbacks, the Flash ADC is in wide use and is considered as a good design option for many systems where high-speed conversions are required. With the availability of inexpensive OpAmps and high-speed programmable digital logic, Flash ADCs will most likely remain as a good design option for many applications.

15.2 VOLTAGE-CONTROLLED OSCILLATORS

In communications circuits, it is often desirable to have the ability to change the oscillation frequency of a oscillator by applying a voltage to the circuit. Oscillators that allow for control of the frequency of oscillation by an applied voltage are called Voltage-Controlled Oscillators (VCO).[2]

---

[1]Precision voltage references are presented in Chapter 14 of this book.
[2]A very simple, low-frequency VCO using multivibrators is presented in Section 13.2.

In practice, the Colpitts oscillator topology is most commonly used to design VCOs in communication electronics. Several phenomena can be exploited to allow voltage control of the oscillation frequency of an electronic oscillator. They included:

- Bias-control to alter amplifier gain and device capacitance.

- Voltage-variable capacitors to alter the circuit resonant frequency.

- Voltage-variable resistors to alter gain.

The first two of these techniques are discussed in this section. Both arrangements alter the reactive LC feedback network of oscillator as a function of an input voltage to alter the oscillation frequency. The use of voltage-variable resistors (e.g., FETs operating in the ohmic region) is not commonly found in VCOs for communication electronics applications: they are, however, used in phase-shift oscillator-type VCOs in low frequency applications.

Bias-Controlled VCO

In Chapter 12, it was shown that the capacitance values of the LC reactive network of a BJT Colpitts oscillator, shown in Figure 15.5, is dependent on the gain of the circuit.

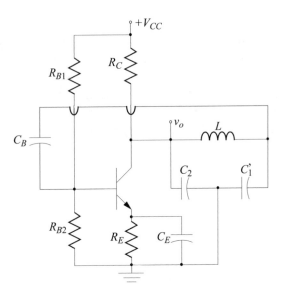

Figure 15.5: Colpitts oscillator circuit.

In order to achieve a particular frequency of oscillation, the capacitance values C_1' and C_2 are highly dependent on the gain of the BJT circuit. The value of $C_1 = C_1' + C_i$, where C_i is the

BJT equivalent input capacitance is given by,

$$\begin{aligned}
C_1 &= C_1' + C_i \\
&= C_1' + C_\pi + C_\mu \left(1 + g_m R\right) \\
&= C_1' + C_\pi + C_\mu \left\{1 + g_{m1} \left[R_{B1} // R_{B2} // R_C // (r_b + r_\pi)\right]\right\}.
\end{aligned} \tag{15.1}$$

The small signal capacitance values associated with the BJT are,

$$C_\mu = \frac{CJC}{\left(1 + \dfrac{V_{CB}}{0.75}\right)^{0.33}}, \tag{15.2}$$

and

$$C_\pi = \left(\frac{g_m}{2\pi f_T}\right) - C_\mu. \tag{15.3}$$

It is evident from Equations (15.1) to (15.3) that the capacitance C_1 is highly dependent of the small-signal mutual conductance, g_m, of the transistor. Since the mutual conductance of a BJT is a function of the bias condition, C_1 is a also a function of the bias current. By changing C_1, the oscillation frequency, $f_o = (1/2\pi) \sqrt{(C_1 + C_2)/(LC_1C_2)}$, can be altered while maintaining a constant value for C_2 and L.

One method of varying the oscillation frequency of a Colpitts oscillator is, therefore, to change the bias condition of the transistor. Figure 15.6 is one implementation of a Colpitts Bias-Controlled VCO. The transistor Q_2 is a constant current source where the bias current is controlled by V_m. The collector current through Q_1 is increased by increasing the control voltage V_m in Q_2. An increase in the bias current of Q_1 will result in a corresponding increase in the capacitor C_1. The capacitor C_E acts as an emitter bypass capacitor for Q_1. The resistor R_{B1} is a bias resistor for Q_1.

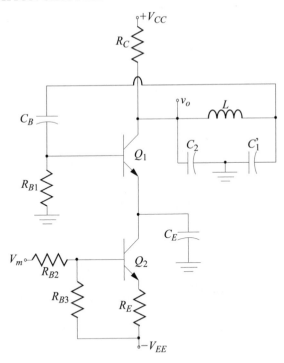

Figure 15.6: Bias-controlled BJT Colpitts VCO.

Example 15.1 Bias-Controlled Colpitts Oscillator Tuning Range

Find the frequency tuning range as a function of the bias control voltage, V_m, for the BJT Colpitts VCO shown. The bias control voltage range is: $-10 < V_m < 10$ V. The BJTs have identical parameters.

The transistor parameters are:

$$\beta_F = 200 \qquad V_A = 200\,\text{V}$$
$$r_b = 30\,\Omega \qquad C_{obo} = 5\,\text{pF}$$
$$f_T = 250\,\text{MHz.}$$

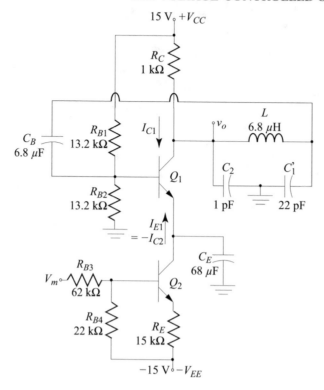

Solution:

Determine the DC condition of the circuit by solving for I_{C1}, V_{CE1}, and V_{CE2} over the tuning voltage range, V_m. The quiescent collector current for Q_1 is:

$$I_{C1} = \left(\frac{\beta_{F1}}{\beta_{F1} + 1} \right) \left(\frac{\beta_{F2}}{\beta_{F2} + 1} \right) \left[\frac{\dfrac{V_m R_{B4}}{R_{B3} + R_{B4}} + V_{EE} - V_{\gamma 1}}{R_E + \dfrac{R_{B4} // R_{B3}}{\beta_{F2} + 1}} \right].$$

The ranges for the collector current of Q_1, V_{CE1}, and V_{CE2} for the bias voltage input of in value from $-10 < V_m < 10$ V are:

$$0.77 \, \text{mA} \quad < I_{C1} \quad < 1.12 \, \text{mA}$$
$$7.08 \, \text{V} \quad < V_{CE1} \quad < 7.43 \, \text{V}$$
$$5.05 \, \text{V} \quad < V_{CE} \quad < 10.24 \, \text{V}.$$

These voltage and collector current ranges confirm that the transistors are in the forward-active region of operation.

The mutual conductance and input resistance of Q_1 are,

$$g_{m1} = \frac{|I_{C1}|}{V_t} = \frac{|I_{C1}|}{0.026},$$

$$R_i = r_{b1} + r_{\pi 1} \approx r_{\pi 1} = \frac{\beta_{F1}}{g_{m1}} = \frac{\beta_{F1} V_t}{|I_{C1}|} = \frac{0.026 \, \beta_{F1}}{|I_{C1}|}.$$

For Q_1 the capacitance, $C_{\mu 1} = C_{obo} = 5\,\text{pF}$. The capacitance, $C_{\pi 1}$, of Q_1 is given by

$$C_{\pi 1} = \left(\frac{g_{m1}}{2\pi \, f_T}\right) - C_{\mu 1}.$$

One of the oscillator capacitors, C_1, is dependent on the bias condition of Q_1,

$$C_1 = C_1' + \frac{g_{m1}}{2\pi \, f_T} - C_{\mu 1} + C_{\mu 1} \left[1 + g_{m1} \left(R_i \, // R_{B1} \, // R_{B2} \, // R_C\right)\right],$$

and the frequency of oscillation is given by,

$$f_o = \frac{1}{2\pi} \sqrt{\frac{C_1 + C_2}{L C_1 C_2}}.$$

Since C_1 is a function of the bias conditions of Q_1, the frequency of oscillation is also a function of bias conditions. Using the equations above, the MathCAD routine shown below is used to find the frequency of oscillation as a function of the bias control voltage.

$$C_{1p} := 22 \cdot 10^{-12} \quad C_2 := 1.0 \cdot 10^{-12} \quad L := 6.8 \cdot 10^{-6} \quad V_{EE} := 15$$
$$\beta_F := 200 \quad C_\mu := 5 \cdot 10^{-12} \quad ft := 250 \cdot 10^6 \quad V_\gamma := 0.7$$
$$i := 0 \ldots 50 \quad Vm_i := -10 + \frac{i}{50} \cdot 20$$

$$I_{C1_i} := \left(\frac{\beta_F}{\beta_F + 1}\right)^2 \cdot \left(\frac{\dfrac{Vm_i \cdot 22}{22 + 62} + V_{EE} - V_\gamma}{15000 + \dfrac{1620}{\beta_F + 1}}\right)$$

$$gm_i := \frac{I_{C1_i}}{0.026} \qquad r\pi_i := \frac{\beta_F \cdot 0.026}{I_{C1_i}}$$

$$C_{1_i} := C_{1p} + \frac{gm_i}{2\pi \cdot ft} - C_\mu + C_\mu \cdot \left[1 + gm_i \cdot \left(\frac{1}{6600} + \frac{1}{1000} + \frac{1}{r\pi_i}\right)^{-1}\right]$$

$$fo_i := \frac{1}{2\pi} \sqrt{\frac{C_{1_i} + C_2}{L \cdot C_{1_i} \cdot C_2}}.$$

The resulting graph is shown below.

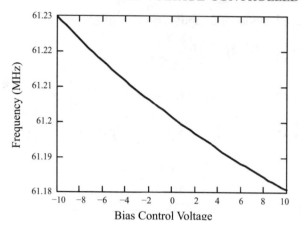

The graph indicates that the VCO tuning range is approximately $61.18\,\text{MHz} < f_o < 61.23\,\text{MHz}$ for $-10 < V_m < 10\,\text{V}$, and is quite linear.

The gain of the circuit is, for all practical purposes, invariant. Therefore, the Barkhausen criterion for oscillation is satisfied over the bias control voltage range.

Varicap VCO

By far the most common method for controlling the oscillation frequency of an oscillator is with a voltage using a varactor diode. The capacitance of a varactor diode (sometimes called a tuning diode, voltage-variable capacitance diode, or varicap) is dependent on the reverse bias voltage across the diode. The depletion capacitance of a diode was presented in Chapter 10 (Book 3) and is repeated here for convenience,

$$C_j = \frac{C_{jo}}{\left(1 - \dfrac{V_d}{\psi_o}\right)^m},$$
(15.4)

where

$C_{jo} \equiv$ small-signal junction capacitance at zero voltage bias (SPICE parameter CJO)
$\psi_o \equiv$ junction built in potential (SPICE parameters VJ)
$m \equiv$ junction grading coefficient, $0.2 < m \le 0.5$ (SPICE parameter MJ).

The symbol for a varactor diode is shown in Figure 15.7.

A common varactor diode is the MV2102.[3] The parameters of interest for this device are $\psi_0 = 0.7266\,\text{V}$, $C_{jo} = 17.88\,\text{pF}$, and $m = 0.424$. A graph of the capacitance as a function of the reverse bias voltage is shown in Figure 15.8.

---

[3]Multisim™ provides a SPICE model for the MV2102 varactor diode as manufactured by Zetex Semiconductors as the FMMV2102 varactor diode.

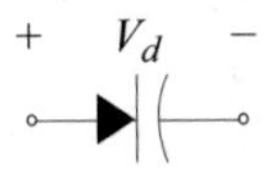

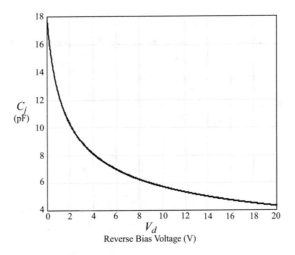

Figure 15.7: Circuit symbol for a varactor diode.

Figure 15.8: Capacitance as a function of reverse bias voltage for an MV2102 varactor diode.

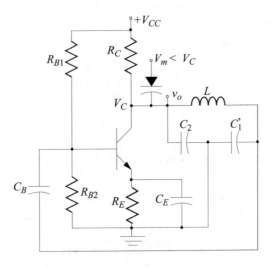

Figure 15.9: Varactor diode-based VCO.

A BJT-based Colpitts VCO can be designed by connecting a varactor diode in parallel (in an AC sense) to C_2 as shown in Figure 15.9. This has the effect of increasing the value of C_2.

Increasing C_2 decreases the oscillation frequency by approximately the square-root of the increase in C_2. Care must be taken so that the voltage at the anode of the diode, V_m, is smaller than the collector voltage, V_C of the BJT so as to not affect the bias conditions of the transistor.

Example 15.2

Tuning voltage of VCO with varactor diode.

For the circuit below, what is the range of the tuning voltage, V_m, so that the VCO will tune over the frequency range of $10.5\,\text{MHz} \le f_o \le 10.7\,\text{MHz}$?

Assume that the transistor parameters are:

$$\beta_F = 200, \quad V_A = 150\,\text{V}, r_b = 30\,\Omega,$$
$$C_\mu = 3\,\text{pF, and } f_T = 250\,\text{MHz}.$$

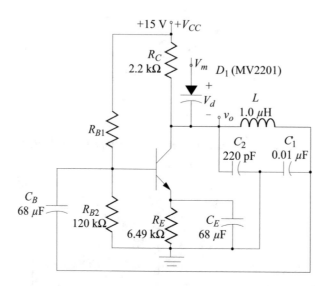

Solution:

The *DC* conditions of this circuit were established in Example 12.2. The capacitance, C_1, was found to be $C_1 = 10.2\,\text{nF}$.

The oscillation frequency of the VCO without the varactor diode is:

$$f_o = \frac{1}{2\pi}\sqrt{\frac{C_1 + C_2}{LC_1C_2}} = \frac{1}{2\pi}\sqrt{\frac{10.2 \times 10^{-9} + 220 \times 10^{-12}}{10^{-6}\left(10.2 \times 10^{-9}\right)\left(220 \times 10^{-12}\right)}} = 10.85\,\text{MHz}.$$

This result implies that for small varicap capacitances, the oscillation frequency approaches 10.85 MHz.

In order to meet the specification for the maximum VCO frequency of 10.7 MHz, $C_2 = 226.15$ pF. The capacitance value of the varicap must add to the physical capacitance in the circuit: 226.15 pF $- 220$ pF $= 6.15$ pF. From the MV2102 characteristic graph of capacitance as a function of reverse bias voltage in Figure 15.8, 6.15 pF corresponds to $V_d = 8.28$ V.

For the minimum VCO frequency specification of 10.5 MHz, $C_2 = 235$ pF. The capacitance value of the varicap must be 235 pF $- 220$ pF $= 15$ pF. From the MV2202 characteristic graph of capacitance as a function of reverse bias voltage in Figure 15.8, 15 pF corresponds to $V_d = 0.37$ V.

Since $I_C = 1$ mA (found in Example 12.2), the *DC* voltage at the collector of the BJT is,

$$V_C = V_{o,DC} = V_{CC} - I_C R_C = 15 - (0.001)(2200) = 12.8 \text{ V.}$$

For a slowly varying voltage, V_m, the reverse bias voltage across the varactor diode is,

$$V_d = V_{o,DC} - V_m.$$

The tuning voltage is then,

$$V_m = V_{o,DC} - V_d.$$

The tuning voltage for a 10.5 MHz oscillation frequency is then,

$$V_m = V_{o,DC} - V_d = 12.8 - 0.37 = 12.43 \text{ V.}$$

The tuning voltage for a 10.7 MHz oscillation frequency is then,

$$V_m = V_{o,DC} - V_d = 12.8 - 8.28 = 4.52 \text{ V.}$$

Therefore, the calculated tuning voltage range is: $4.52 \text{ V} < V_m < 12.43 \text{ V}$.

15.3   MIXERS

A mixer uses the non-linearity of a device to produce intermodulation[4] products. In most applications, a mixer is used to generate the difference frequency between the input signal (commonly called the *RF* in radio frequency applications) and the local oscillator (called the *LO*; this is just a stable oscillator circuit). Consider the output from a device with non-linear characteristics,

$$v_o = V_{DC} + a_1 v_i + a_2 v_i^2 + a_3 v_i^3 + \cdots. \tag{15.5}$$

When two sinusoids are mixed, the input voltage can be represented as,

$$v_i = X_1 \cos \omega_{RF} t + X_2 \cos \omega_{LO} t. \tag{15.6}$$

---

[4]Intermodulation distortion was discussed in Chapter 7 (Book 2). Intermodulation products are those frequency components that are produced at the output of a device with non-linear characteristics with two inputs of different frequency.

Substituting Equation (15.6) into Equation (15.5) yields,

$$v_O = V_{DC} + a_1 (X_1 \cos \omega_{RF}t + X_2 \cos \omega_{LO}t) + a_2(X_1 \cos \omega_{RF}t + X_2 \cos \omega_{LO}t)^2$$
$$+ a_3(X_1 \cos \omega_{RF}t + X_2 \cos \omega_{LO}t)^3 + \cdots . \tag{15.7}$$

By using trigonometric identities, Equation (15.7) yields the magnitude and frequency components of the output signal that includes the difference frequency $\omega_{RF} - \omega_{LO}$,

$$v_o = \cdots + a_2 X_1 X_2 \cos (\omega_{RF} - \omega_{LO}) t + \cdots . \tag{15.8}$$

This desired difference frequency can be recovered using a bandpass filter.

In typical applications, two signals, *RF* and *LO* are applied to the mixer, from which the difference (or other) intermodulation distortion product is selected. In a receiver, the known signal is provided by the local oscillator; the unknown signal is the *RF*. These two signals are mixed and filtered. This process of intentionally creating intermodulation products is called *heterodyning*. The output of the mixing and filtering process in a receiver is the intermediate frequency (*IF*).

Heterodyning has many advantages. One of the advantages heterodyning in a receiver lies in the reduction of operating frequency from the *RF*. That is, the *RF* signal is typically a very high frequency signal. By mixing the *RF* with an *LO* of a specified frequency, the *IF* is significantly lower in frequency than the *RF*, making circuit design less complex.

The symbol for a mixer is shown in Figure 15.10.

Figure 15.10: Symbol for a mixer.

15.3.1 BJT MIXERS

Typically, fundamental building block mixer designs include small-signal *RF* amplifiers. Therefore, an understanding of small-signal *RF* amplifier design is critical in the design of mixers. A small-signal BJT *RF* amplifier will be used to demonstrate this design technique.

Small-Signal *RF* Amplifier Design
As in all electronic circuit design, the design of small-signal *RF* amplifiers involves a great number of choices and exercise in good judgment. Logically derived rules-of-thumb may be applied to narrow the options available to the designer.

For example, a single-transistor *RF* amplifier may utilize a self-biasing scheme for maximum temperature and device variance stability, as shown in Figure 15.11.

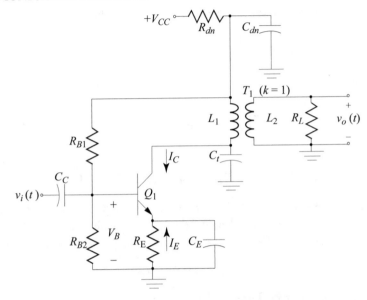

Figure 15.11: Small-signal *RF* amplifier—self-bias configuration.

The design process for this amplifier is as follows:

- Establish stable quiescent conditions

- Optimize collector bias currents for maximum output power

- Decouple the power supplies

- Establish the value of the *DC* blocking capacitor

- Determine the tuning capacitor value to select the desired signal product

- Determine the effective load at the BJT collector using the transformer Q

For quiescent point stability defined by a 1% (or less) change in collector current for a 10% change in β_F, the biasing rule-of-thumb is,

$$\frac{R_B}{R_E} \leq \frac{\beta_F}{9} - 1, \tag{15.9}$$

where

$$R_B = R_{B1} // R_{B2}. \tag{15.10}$$

Another rule of thumb sets the voltage across R_E to 10% of V_{CC},

$$R_E = -\frac{V_{CC}}{10I_E}. \tag{15.11}$$

A convenient value for R_{B2} is,

$$R_{B2} = \frac{V_B}{10I_B} = \frac{V_B \beta_F}{10I_C}, \tag{15.12}$$

where

$$V_B = -I_E R_E + V_{BE}, \quad \text{and} \quad V_{BE} \approx 0.7\,\text{V}.$$

Optimum collector bias current is based on the maximum output (RF) power P_o and power supply voltage V_{CC}. A transformer-coupled class A amplifier can achieve a maximum efficiency of 50% at full output voltage swing. If V_E is 10% of V_{CC}, then the power supply is providing

$$P_{DC} = 0.9V_{CC}I_C, \tag{15.13}$$

to the collector of the transistor. For a 50% efficiency,

$$P_{DC} = \frac{P_O}{\eta} = 2P_O, \tag{15.14}$$

such that

$$I_C = \frac{2P_O}{0.9V_{CC}}. \tag{15.15}$$

For example, if $P_O = 10\,\text{mW}$ and $V_{CC} = 12\,\text{V}$, $I_C = 20\,\text{mW}/(0.9 \times 12\,\text{V}) = 1.85\,\text{mA}$. The emitter resistance is then $R_E = 0.1(12\,\text{V})/1.85\,\text{mA} = 649\,\Omega$.

R_{dn} and C_{dn} form a lowpass filter called a decoupling network which provides an AC low impedance point between the collector and base. The decoupling network also isolates the amplifier from the power supply rail from possible feedback to other amplifiers. RF chokes are not recommended as a replacement for R_{dn} because of potential resonance effects with C_{dn}. A value of $R_{dn} = 100\,\Omega$ is typical for low power amplifiers and will result in only a few tenths of volts coupled away from the amplifier. C_{dn} is a bypass capacitor that is chosen for a reactance of one order of magnitude less than the resistance. For example, if the amplifier is used at 455 kHz and $R_{dn} = 100\,\Omega$, then $X_C \leq 10\,\Omega$ so that $C_{dn} \geq 0.035\,\mu\text{F}$. C_{dn} is in series with C_E (AC path). Because C_E typically has a reactance of a few ohms, it is appropriate to make $C_{dn} = C_E$.

The emitter bypass capacitor C_E provides an AC low impedance for the transistor emitter. It is typically sufficient in RF circuit design (low impedance circuits) to make X_{CE} an order of magnitude less than the inverse of the transconductance of the transistor,

$$X_{CE} = \frac{1}{10g_m} = \frac{V_t}{10I_C}. \tag{15.16}$$

The value of the *DC* blocking capacitor C_C is determined in the same way as a bypass except that its reactance should be an order of magnitude less that the amplifier input impedance: that is, $X_{CC} = Z_{in}/10$, where Z_{in} is the amplifier input impedance. The Z_{in} of an *RF* amplifier must include all input capacitances.

The capacitor C_t is a tuning capacitance to tune to the proper Q and bandwidth at the *IF* frequency,

$$C_t = \frac{1}{(2\pi f_{IF})^2 L_1},$$ (15.17)

where f_{IF} is the *IF* frequency.

At the output, the effective quality factor, Q_{eff} of the transformer must be determined to determine the load reflected onto the collector of the transistor. First, the reactance of the primary of the transformer X_{L1} is determined. Knowing the desired bandwidth of $f_{IF}, BW_{IF}, Q_{eff} = f_{IF}/BW_{eff}$: then $R'_{L1} = X_{L1} \times Q_{eff}$ and

$$n_p/n_s = \sqrt{R'_{L1/R_L}}.$$

Active Mixer Design

The design of an active mixer is the same as that of the small-signal *RF* amplifier. The only difference is an additional input for the local oscillator (*LO*) as shown in Figure 15.12.

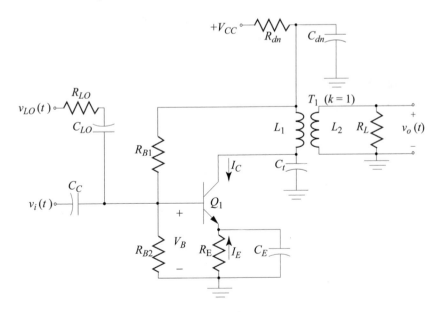

Figure 15.12: Active mixer.

The coupling capacitor C_{LO} is,

$$C_{LO} = \frac{1}{2\pi f_{LO} R_{LO}}, \tag{15.18}$$

where R_{LO} is the *LO* source resistance.

15.3.2 DUAL GATE FET MIXER

A dual-gate FET is an *n*-channel depletion type FET (commonly a GaAs MESFET) with two independently insulated gate terminals. MESFETs operate essentially like a MOSFET, but can be used at very high frequencies, on the order of several GHz. These FETs have a series arrangement of two separate channels, with each channels having independent gate control. The circuit symbol for the *n*-channel depletion type dual-gate MESFET is shown in Figure 15.13a, with the simplified symbol in Figure 15.13b. Dual-gate MESFETs are used at very high frequencies and are commonly used in mixer, modulator, and Automatic Gain Control (AGC) circuits. When Gate 2 is at AC ground, the dual-gate FET may be represented as a common-source, common-gate (CS-CG) pair, referred to as the cascode circuit, shown in Figure 15.14.

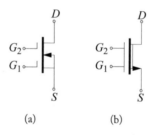

Figure 15.13: (a) Dual-gate *n*-channel depletion MESFET symbol; (b) Simplified dual-gate MESFET circuit symbol.

Figure 15.14: Dual-gate *n*-channel depletion MESFET equivalent representation.

In mixer/modulator applications, the *LO* is applied to Gate 1, and the output taken from the drain (common-source configuration). The baseband signal is applied to Gate 2 to modulate the *LO* signal.

A dual-gate *n*-channel depletion MESFET mixer makes use of the isolated gates for good isolation between the *LO* and the baseband is shown in Figure 15.15. The *IF* output is transformer coupled for impedance matching and isolation.

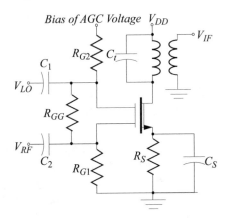

Figure 15.15: Dual-gate *n*-channel depletion type MESFET mixer.

15.4 PHASE-LOCK LOOPS

The phase-lock loop (PLL) is used to track the phase and frequency of a signal. It is often used in the receiver in both amplitude modulated (AM) and frequency modulated (FM) systems. In addition to AM and FM receiver applications, PLLs are used in control applications, such as in a compact disk (CD) player, to track the rotational speed of the CD.

The PLL is a feedback system as shown in Figure 15.16. In the PLL, the feedback signal is intended to follow the phase and frequency of the input signal. However, in the case where the input and feedback signals are not equal, the difference between the two, called the error signal, will change the feedback signal. The feedback signal will continue to change and until it again matches the input signal. The feedback quantity that is compared with the input signal is a generalized phase angle $\theta(t)$. A generalized phase angle is composed of a frequency and phase component: that is,

$$\theta(t) = \omega_c t + \phi_c(t), \tag{15.19}$$

where ω_c is a set frequency, often called the carrier frequency in communication systems, and $\phi_c(t)$ is a time varying phase.

The frequency of the output signal of the VCO is adjusted until its generalized phase angle is close to that of the input signal of the PLL. When the generalized phase angles of the input

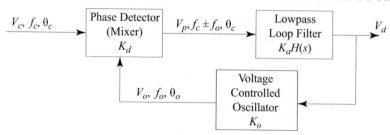

Figure 15.16: Block diagram of a phase-lock loop.

signal and the output of the VCO are nearly identical, the two signals are synchronized. There is, however, a constant phase difference between the two signals due to signal delay through the PLL.

Although several different analog and digital implementations of the PLL are widely used, one of the most common configurations uses a mixer as a phase comparator, a loop filter with a response $K_a H(s)$, and a VCO. The error or difference voltage, V_d, is the output of the loop filter which is the controlling voltage to the VCO. For a quiescent VCO frequency of ω_c, the error voltage $V_d = 0$, and the loop is said to be in lock.

Phase Detector

When the PLL is in lock, the output voltage of the phase detector is the difference frequency signal with phase difference, or static phase error, $\phi_e = \phi_c - \phi_o$. If the input signal to the mixer phase detector is $v_c(t) = V_c \sin(\omega_c t + \phi_c)$, the reference signal from the VCO is $v_o(t) = V_o \sin(\omega_c t + \phi)$, and $\omega_c = \omega_o$, then the output signal of the mixer phase detector is,

$$V_p(t) = v_c(t) v_o(t) = \frac{V_c V_o}{2} K \cos \phi_e - \frac{V_c V_o}{2} K \cos(2\omega_o t + \phi_e), \qquad (15.20)$$

where K is the mixer gain.

Since the lowpass loop filter eliminates the second harmonic term from the output of the mixer phase detector, only the first term of Equation (15.20) is considered: that is,

$$V_p = \frac{V_c V_o}{2} K \cos \phi_d. \qquad (15.21)$$

For the case where the input frequency, f_s, is equal to the free-running quiescent frequency of the VCO, f_o, $\phi_e = \pi/2$ and the difference voltage $V_d = 0$. Therefore, the output voltage of the phase detector is also zero. The error signal is proportional to phase differences about 90°. For small changes in phase $\Delta \phi_e$,

$$\phi_e \approx \frac{\pi}{2} + \Delta \phi_e, \qquad (15.22)$$

and the mixer phase detector output is,

$$V_p = \frac{V_c V_o}{2} K \cos\left(\frac{\pi}{2} + \Delta\phi_e\right) = \frac{V_c V_o}{2} K \sin(\Delta\phi_e)$$

$$\approx \frac{V_c V_o}{2} K \Delta\phi_e. \tag{15.23}$$

For small phase perturbation, $\Delta\phi_e$,

$$V_p \approx \frac{V_c V_o K}{2} \Delta\phi_e, \tag{15.24}$$

under the assumption that

$$V_p = K_d (\phi_c - \phi_o), \tag{15.25}$$

where K_d is the phase detector scale factor defined as,

$$K_d = \frac{V_c V_o K}{2}.$$

The gain of each of the components of the PLL must be defined in order to find the closed loop transfer function. When the loop is in lock, the gain factor of the phase detector is,

$$K_d = \frac{V_p}{\sin \Delta\phi_e} \text{V/rad.} \tag{15.26}$$

In PLLs, the phase $\Delta\phi_e$ is usually designed to be small so that a pulse of noise will not throw the loop out of lock.

Loop Filter

The lowpass loop filter can be passive or active. When passive filters are used, an amplifier with gain K_a is usually required to increase the amplitude of the filtered difference signal. Two lowpass passive filters are shown in Figure 15.17. A first or second order Butterworth lowpass filter may be used as the active loop filter in the PLL.

For the simple lowpass filter without gain in Figure 15.17a, the transfer function is,

$$H(s) = \frac{1}{1 + s R_1 C}. \tag{15.27}$$

For the lag-lead loop filter without gain shown in Figure 15.17b, the transfer function is,

$$H(s) = \frac{1 + s R_2 C}{1 + s (R_1 + R_2) C}. \tag{15.28}$$

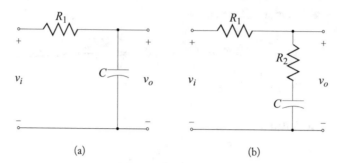

Figure 15.17: Two passive lowpass loop filter configurations.

Voltage-Controlled Oscillator

The output frequency of the VCO is expressed as,

$$f_o = f_f + \frac{K_o}{2\pi} v_d, \tag{15.29}$$

where

f_o is the oscillator output frequency in Hertz,

f_f is the quiescent frequency of the VCO in Hertz, and

K_o is the VCO voltage to frequency relationship in radians per volt (rad/V).

Equation (15.29) can be written in terms of radian frequency,

$$\omega_o = \omega_f + K_o v_d. \tag{15.30}$$

The total generalized phase angle of the output of the VCO is,

$$\theta(t) = \int_0^t (\omega_f + \Delta\omega)\, dt = \omega_f t + \phi_o(t), \tag{15.31}$$

where $\Delta\omega$ is the frequency deviation from ω_f.

Therefore, the phase term is defined as,

$$\phi_o(t) = \int_0^t \Delta\omega\, dt, \tag{15.32}$$

or

$$\frac{d\phi_o(t)}{dt} = \Delta\omega = K_o v_d. \tag{15.33}$$

The difference voltage, v_d, is a *DC* voltage when the loop is in lock. When not the PLL is not in lock, v_d is a voltage corresponding to the difference frequency $(f_c - f_o)$ that draws the VCO in to synchronization with the input signal. When the PLL is in lock, the VCO output frequency equals that of the input signal. However, there is a phase difference detected by the phase detector between the VCO and input signals. This difference is called the static phase error, ϕ_e. The static phase error is used to maintain the necessary control voltage on the VCO to maintain the required frequency to keep the PLL in lock.

In the s-domain, Equation (15.33) is,

$$\Phi_o(s) = K_o \frac{V_d(s)}{s}, \tag{15.34}$$

which clearly shows that the VCO performs as an integrator for phase errors. As an integrator, the VCO helps maintain PLL lock through momentary disturbances.

Closed-Loop Transfer Function

The closed loop transfer function of a PLL is found by determining the ratio $\Phi_o(s)/\Phi_c(s)$, where $\Phi_o(s)$ and $\Phi_c(s)$ are frequency domain representations of $\phi_o(t)$ and $\phi_c(t)$, respectively. This quantity provides a measure of the loop response to changes in the input phase or frequency. The VCO input signal is,

$$V_d(s) = V_e(s) H(s) K_a, \tag{15.35}$$

and

$$V_e(s) = K_d \Phi_e(s). \tag{15.36}$$

The *DC* loop gain is,

$$K_v = K_d K_a K_o = \left(\frac{V_e(s)}{\Phi_e(s)}\right)\left(\frac{V_d(s)}{V_e(s)}\right)\left(\frac{\Delta\omega}{V_d(s)}\right) = \frac{\Delta\omega}{\Phi_e(s)}. \tag{15.37}$$

Substituting Equations (15.35) through (15.37) into (15.34) yields,

$$\Phi_o(s) = K_v \Phi_e(s) \frac{H(s)}{s}. \tag{15.38}$$

The open-loop transfer function is defined as,

$$P(s) = \frac{\Phi_o(s)}{\Phi_e(s)} = K_v \frac{H(s)}{s}, \tag{15.39}$$

and the closed-loop transfer function is,

$$F(s) = \frac{\Phi_o(s)}{\Phi_c(s)} = \frac{P(s)}{1 + P(s)} = \frac{K_v H(s)}{s + K_v H(s)}. \tag{15.40}$$

From Equation (15.40), it is apparent that the transfer function of the loop filter is a major factor in determining the loop performance. When the filter bandwidth is reduced, the response time to changes in phase or frequency is increased, and helps maintain loop lock against momentary disturbances of the input signal.

Other types of phase detectors are also used in commercial PLLs. The type of phase detector circuit selected depends on many factors including, cost, size, speed, noise performance, and manufacturability. Mixer phase detectors are most commonly used in applications where little VCO frequency deviation from the free-running state is required. For high-speed performance, digital phase detectors using Emitter-Coupled Logic (ECL) circuitry are usually preferred and are commercially available in single chip PLL packages.

Example 15.3 Loop Characteristics
For the PLL shown below, determine:

1. Gain of the OpAmp amplifier

2. Loop gain in units of 1/seconds and in dBs at $\omega = 1$ rad/s

3. VCO output frequency when the PLL is phase-locked

4. Static phase error, ϕ_e, and V_d when the PLL is phase-locked.

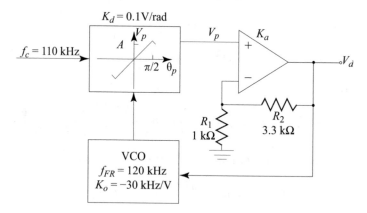

Solution:

(a) Gain of the OpAmp amplifier.

The OpAmp amplifier is in the non-inverting configuration. Therefore, the gain of the amplifier is,

$$K_a = 1 + \frac{R_2}{R_1} = 1 + \frac{3.3\,k}{1\,k} = 4.3.$$

(b) Loop gain in units of 1/seconds and in dBs at $\omega = 1$ rad/s.

$$K_v = K_d K_a K_o = (0.1 \text{ rad/V}) (4.3) (-30 \text{ kHz/rad}) = -12.9 \text{ kHz/V}.$$

Converting -12.9 kHz/V to units of 1/s,

$$K_v = (-12.9 \text{ kcycles/s} - \text{rad}) (2\pi \text{ rad/cycle}) = -81,053 \text{ s}^{-1}.$$

Expressed in dB,

$$K_{v,\text{dB}} = 20 \log (|K_v|) = 20 \log (81053) = 98.2 \text{ dB at 1 rad/s}.$$

(c) VCO output frequency when the PLL is phase-locked.

When the PLL is locked, the VCO frequency is, by definition, $f_o = f_i = 110$ kHz. Only a phase difference can exist between the input signal and the VCO output at the phase detector. This phase difference is called the static phase error that yields a difference voltage output from the loop filter which provides sufficient frequency control voltage to the VCO to maintain frequency lock.

(d) Static phase error, ϕ_e, and V_d when the PLL is phase-locked.

The VCO control voltage, V_d, is,

$$V_d = \frac{f_c - f_o}{K_o} = \frac{(110 \text{ kHz} - 120 \text{ kHz})}{-30 \text{ kHz/V}} = 0.33 \text{ V}.$$

Since the gain of the OpAmp amplifier is 4.3, the output of the phase detector/filter, V_p, is

$$V_p = \frac{0.33 \text{ V}}{4.3} = 0.077 \text{ V}.$$

The static phase error is,

$$\phi_e = \frac{V_p}{k_d} = \frac{0.077 \text{ V}}{0.1 \text{ V/rad}} = 0.77 \text{ rad}.$$

Hold-In Range

The range of frequencies over which the loop maintains lock is called the *hold-in* range. For a PLL where the amplifier does not saturate and the VCO has a wide frequency range, the phase detector characteristic limits the hold-in range. As static phase error increases due to increasing input frequency, f_c, a limit for the output of the phase detector is reached beyond which the phase detector cannot supply additional corrective control voltage to the VCO. If the phase detector cannot produce more than $V_{p,\text{max}}$, the total range of the phase detector output is $\pm V_{p,\text{max}}$, for a total range the static phase error $\phi_e = \phi_c - \phi_o$ is π radians. The hold-in frequency is the minimum to maximum input frequency range $f_{c,\max} - f_{c,\min} = \Delta f_H$,

$$\Delta f_H = \frac{K_v}{4\pi}, \tag{15.41}$$

where K_v is in radians/second.

15.5   ACTIVE AND PASSIVE FILTER DESIGN

Signal filtering is often central to the design of many communication subsystems. The isolation or elimination of information contained in frequency ranges is of critical importance. In simple amplitude modulation (AM) radio receivers, for example, the user selects one radio station using a bandpass filter techniques. Other radio stations occupying frequencies close to the selected radio station are eliminated. The standard AM broadcast band in North America is from 535 kHz to 1605 kHz.

In Chapter 9 (Book 3), active filter concepts using OpAmps were introduced. One of the advantages of using active filters included the addition of some gain. However, due to their limited gain-bandwidth product, active filters using OpAmps see little use in communication system design.

The two types of frequency selective circuit configurations most commonly used in communication systems are the passive LC filter (low, high, and bandpass responses) and the tuned amplifier (bandpass response). In this section, an overview of these two types of frequency selective circuit configurations for use in bandpass applications is provided.

15.5.1 PASSIVE LC FILTERS

A capacitor coupled low-pass filter configuration is shown in Figure 15.18. This type of filter consists of L and C components connected in a "ladder" network with component values, g_n, determined by the polynomial coefficients for the desired type of filter (i.e., Butterworth or Chebyshev polynomials) and the required system input and output impedances. For this section, the system input and output impedances are assumed to be the same value (a typical value for many RF applications is 50 Ω). Similarly, the g_n values are normalized to these identical input and output resistances. The determination of component values begins with the specifications and procedures outlined in Chapter 9 (Book 3) and specified source and load resistances.
For simplicity, Chebyshev filters are commonly chosen to be odd-order filters: this choice results in identical input and output resistances. Therefore, the first and last filter normalized component values, g_o and g_{N+1}, are equal to 1 for both Butterworth and these odd-order Chebyshev filters.

By setting the first and last normalized component values, g_o and g_{N+1} respectively, equal to 1 (equal input and output resistance R), the intermediate inductor and capacitor values for the filters are found to be:

$$C = \frac{g_n}{2\pi f_c R} \quad \text{and} \quad L = \frac{g_n R}{2\pi f_c},$$

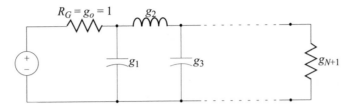

Figure 15.18: Low pass filter "ladder" network.

where

f_c is the cutoff frequency (Hz),

R is the input and output resistance (commonly $50\,\Omega$),

g_n are the Butterworth or Chebyshev polynomial values.

For Butterworth filters, the polynomial values, g_n, are obtained by doubling the factored polynomial damping coefficients (Table 9.2 (Book 3))—a first order factor results in a g value of 2. Determination of the Chebyshev filter values is a more complex operation: a selection of values is listed in Table 15.1.

High-pass and band-pass filter designs are achieved using techniques similar to those described above and in Chapter 9 (Book 3) by replacing the normalized prototype low-pass filter components with those shown in Table 15.2.

For band-pass and band-stop values, the bandwidth BW is

$$BW = \omega_U - \omega_L,$$

where

ω_U is the upper cutoff frequency of the band-pass or band-stop filter

ω_L is the lower cutoff frequency of the band-pass or band-stop filter.

Table 15.1: Normalized Chebyshev filter values

N	g_1	g_2	g_3	g_4	g_5	g_6	g_7	g_8	g_9	g_{10}	g_{11}
Low-Pass Chebyshev Filter: 0.5 dB Ripple											
1	0.6986	1.0000									
2	1.4029	0.7071	1.9841								
3	1.5963	1.0967	1.5963	1.0000							
4	1.6703	1.1926	2.3661	9.8419	1.9841						
5	1.7058	1.2296	2.5408	1.2296	1.7058	1.0000					
6	1.7254	1.2479	2.6064	1.3137	2.4758	0.8696	1.9841				
7	1.7372	1.2583	2.6381	1.3444	2.6381	1.2583	1.7372	1.0000			
8	1.7451	1.2647	2.6564	1.3590	2.6964	1.3389	2.5093	0.8796	1.9841		
9	1.7504	1.2690	2.6678	1.3673	2.7939	1.3673	2.6678	1.2690	1.7504	1.0000	
10	1.7543	1.2721	2.6754	1.3275	2.7392	1.3806	2.7231	1.3485	2.5239	0.8842	1.9841

N	g_1	g_2	g_3	g_4	g_5	g_6	g_7	g_8	g_9	g_{10}	g_{11}
Low-Pass Chebyshev Filter: 3 dB Ripple											
1	1.9953	1.0000									
2	3.1013	0.5339	5.8095								
3	3.3487	0.7117	3.3487	1.0000							
4	3.4389	0.7483	4.3471	0.5920	5.8095						
5	3.4817	0.7619	4.5381	0.7618	3.4817	1.0000					
6	3.5045	0.7685	4.6061	0.7929	4.4641	0.6033	5.8095				
7	3.5182	0.7723	4.6386	0.8039	4.6386	0.7723	3.5182	1.0000			
8	3.5277	0.7745	4.6575	0.8089	4.6990	0.8018	4.4990	0.6073	5.8095		
9	3.5340	0.7760	4.6692	0.8118	4.7272	0.8118	4.6692	0.7760	3.5340	1.0000	
10	3.5384	0.7771	4.6768	0.8136	4.7425	0.8164	4.7260	0.8051	4.5142	0.6091	5.8095

Table 15.2: Component transformation from low-pass values

Low-pass Prototype	Low-pass	High-pass	Bandpass	Bandstop
$L = g_k$	$\dfrac{L}{\omega_c}$	$\dfrac{1}{\omega_c L}$	$\dfrac{L}{BW}$ $\dfrac{BW}{\omega_o^2 L}$	$\dfrac{1}{(BW)L}$ $\dfrac{(BW)L}{\omega_o^2}$
$C = g_k$	$\dfrac{C}{\omega_c}$	$\dfrac{1}{\omega_c C}$	$\dfrac{C}{BW}$ $\dfrac{BW}{\omega_o^2 C}$	$\dfrac{1}{(BW)C}$ $\dfrac{(BW)C}{\omega_o^2}$

Example 15.4   Design Example

Design a Butterworth low-pass filter with 3 dB cut-off frequency of 900 MHz and a stop band attenuation of 22 dB at 1.8 GHz. The source and load impedances are 50 Ω.

Solution:

The first task is to find the required order of the filter which fulfills the specifications. Using the technique developed in Chapter 9 (Book 3), the order is found to be:

$$10^{22/20} = \sqrt{1 + \left(\frac{1800}{900}\right)^{2N}} \quad \text{or} \quad N = \frac{\log\left(10^{22/10} - 1\right)}{2\log(1800/900)} = 3.65.$$

Since filters must be integer order, a 4th order filter is necessary.

The normalized component values for the low-pass filter are determined from the fourth-order Butterworth damping coefficients as tabulated in Table 9.2 (Book 3). Each normalized component value is double a Butterworth damping coefficient:

$$g_1 = g_4 = 2 \times 0.3827 = 0.7654$$
$$g_2 = g_3 = 2 \times 0.9239 = 1.8478.$$

The value of $g_5 = 1$ is the normalized load resistance that is equal to the normalized source resistance, $g_0 = 1$.
The filter component values are computed to be:

$$R_0 = 1(50) \qquad \text{Source Resistance} \qquad = 50\,\Omega$$
$$C_1 = 0.7654/(2\pi f_c(50)) \qquad \text{Shunt Capacitor, } C_1 \qquad = 2.71\,\text{pF}$$
$$L_2 = 1.8478(50)/(2\pi f_c) \qquad \text{Series Inductor, } L_2 \qquad = 16.33\,\text{nH}$$
$$C_3 = 1.8478/(2\pi f_c(50)) \qquad \text{Shunt Capacitor, } C_3 \qquad = 6.54\,\text{pF}$$
$$L_4 = 0.7654(50)/(2\pi f_c) \qquad \text{Series Inductor, } L_4 \qquad = 6.77\,\text{nH}$$
$$R_L = 1(50) \qquad \text{Load Resistance} \qquad = 50\,\Omega.$$

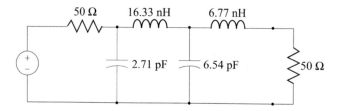

The SPICE simulation normalized frequency response is shown in Figure 15.19.

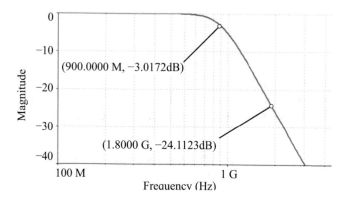

Figure 15.19: Low pass filter response for Example 15.4.

15.5.2 TUNED AMPLIFIERS

Tuned amplifiers are designed to amplify a narrow band of frequencies centered about the center frequency f_o. They can be used as amplifying bandpass filters in contrast to the passive LC filters previously discussed. Unlike wideband amplifiers that were presented in previous chapters, the design of tuned amplifiers requires careful design to avoid oscillations due to the reactive load and the internal reactive feedback elements of the transistor. Since tuned amplifiers only operate over a narrow band of frequencies ($\pm 5\%$ of the center frequency), the two-port parameters of the active device at the center frequency are used in the design.

The simplest, and most common, form of tuned amplifier is the single-tuned amplifier. The single-tuned amplifier is designed using a FET, BJT, or OpAmp. Naturally, the OpAmp implementation suffers from bandwidth limitations due to the OpAmp gain-bandwidth product.

The one common denominator found in different types of single-tuned amplifiers is the use of a passive parallel resonant circuit shown in Figure 15.20a using a non-ideal finite-Q (unloaded) inductor and capacitor. The non-ideal inductor can be modeled as an ideal inductor with a series resistor, r, as shown in Figure 15.20b.

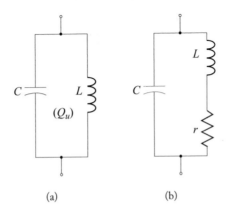

(a)   (b)

Figure 15.20: Passive parallel resonant circuit.

The relationship between the unloaded Q_u of the inductor, resonant frequency, and the inherent series resistance, r, is,

$$Q_u = \frac{\omega_o L}{r}.$$

(15.42)

The admittance of the circuit is,

$$Y = j\omega C + \frac{1}{r + j\omega L}$$

$$= \frac{r}{r^2 + \omega^2 L^2} + j\left[\omega C - \frac{\omega L}{r^2 + \omega^2 L^2}\right].$$

(15.43)

At the resonant frequency, $\omega = \omega_o$ and the imaginary component goes to zero. Therefore,

$$\omega_o C = \frac{\omega_o L}{r^2 + \omega_o^2 L^2}.$$

(15.44)

Solving for ω_o yields the expression for the resonant frequency in terms of the passive components,

$$\omega_o = \sqrt{\frac{1}{LC} - \left(\frac{r}{L}\right)^2}. \tag{15.45}$$

Using Equation (15.42), Equation (15.45) can be re-written in terms of the unloaded Q of the inductor,

$$\omega_o = \sqrt{\frac{1}{LC}\left(\frac{Q_u^2}{Q_u^2 + 1}\right)}. \tag{15.46}$$

A finite Q inductor has the effect of reducing the resonant natural frequency of an ideal LC circuit by a factor of $Q_u^2/(Q_u^2 + 1)$. Therefore, the series RL branch in Figure 15.20b can be replaced with a resistor R_p and L_p in parallel, creating a parallel RLC circuit as shown in Figure 15.21.

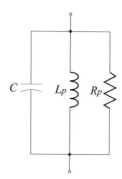

Figure 15.21: Parallel equivalent circuit of LC circuit using non-ideal inductors.

The parallel equivalent circuit parameters have the values,

$$R_p = r\left(Q_u^2 + 1\right), \tag{15.47}$$

and

$$L_p = L_s\left(\frac{Q_u^2 + 1}{Q_u^2}\right). \tag{15.48}$$

If $r \ll \omega L$ then,

$$R_p = \frac{\omega^2 L}{r} = \omega L Q_u = r Q_u^2. \tag{15.49}$$

The impedance of the parallel equivalent circuit is,

$$Z(\omega) = \frac{R_p}{1 + jQ_u\left(\dfrac{\omega}{\omega_o} - \dfrac{\omega_o}{\omega}\right)}. \tag{15.50}$$

A single-tuned amplifier uses the parallel LC resonant circuit (commonly called the tank circuit) in place of a load resistor. The small resistance inherent in the non-ideal inductor must be taken in to account when biasing the circuit. Since the impedance is highest at the resonant frequency, f_o, the gain of the circuit peaks at that frequency. For frequencies far from f_o, the load impedance is small which has the effect of reducing gain.

FET Single-Tuned Amplifier

A simple single-tuned NMOSFET amplifier is shown in Figure 15.22. The resonant components of this tuned amplifier are $R_D//r_d$, $C_T//C_o$, and L_T, where C_o is the effective output capacitance of the NMOSFET. The inductor is non-ideal with a unloaded Q factor specified by Q_u.

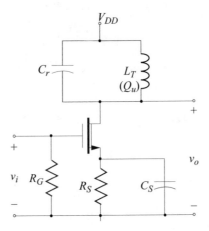

Figure 15.22: Single-tuned NMOSFET amplifier.

Using the parallel equivalent model of an LC tank circuit shown in Figure 15.21, the small-signal model of the single-tuned NMOSFET amplifier of Figure 15.22 is shown in Figure 15.23. The resistance R_p is the resistance associated with inductor.

The voltage transfer characteristic of the circuit is,

$$A_v = \frac{v_o}{v_i} = -\frac{g_m}{G + sC + \dfrac{1}{sL_p}} = -\frac{g_m}{C}\left(\frac{s}{s^2 + s\left(\dfrac{G}{C}\right) + \dfrac{1}{L_pC}}\right), \tag{15.51}$$

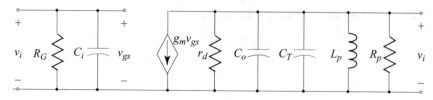

Figure 15.23: Small-signal equivalent circuit of an NMOSFET single-tuned amplifier.

where

$$G = \frac{1}{R_p // r_d}, \quad \text{and} \quad C = C_o + C_T.$$

The magnitude of the gain is,

$$|A_v(j\omega)| = \frac{g_m}{C} \left[\frac{\omega}{\sqrt{\frac{G}{C}\omega^2 + \left(\frac{1}{L_p C} - \omega^2\right)^2}} \right]. \tag{15.52}$$

Equation (15.52) is maximum-valued when the circuit is operating at the resonant frequency, ω_o, with bandwidth $\omega_{3\,\text{dB}}$ defined as,

$$\omega_o = \frac{1}{\sqrt{L_p C}} \quad \text{and} \quad \omega_{3\,\text{dB}} = \frac{1}{RC}. \tag{15.53}$$

The gain at resonance is,

$$|A_v(j\omega_o)| = -\frac{g_m}{G} = -g_m(r_d // R_p). \tag{15.54}$$

BJT Single-Tuned Amplifier
A BJT tuned amplifier is shown in Figure 15.24. Note that the resonant circuit is at the input of the transistor amplifier V_{CC}.
The current gain of the circuit shown in Figure 15.24 is,

$$A_v = \frac{i_o}{i_i} = -\frac{g_m}{G + sC + \frac{1}{sL_p}} = -\frac{g_m}{1 + sRC + \frac{R}{sL_p}}, \tag{15.55}$$

where

$$C = C_T + C_i, \quad R = R_{B1} // R_{B2} // R_P, \quad G = 1/R.$$

Equation (15.55) is of the same form as Equation (15.51). Therefore, the expression for the resonant frequency and bandwidth of the circuit is identical to that of the FET tuned amplifier.

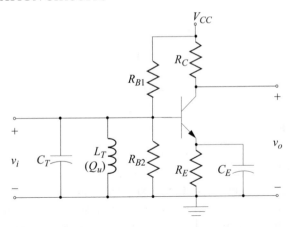

Figure 15.24: BJT tuned amplifier.

15.6   MODULATOR/DEMODULATOR DESIGN

Modulation is the process of combining information with a separate waveform to allow efficient transmission. Demodulation is the process of recovering information from waveform that has been altered to carry that information. There are a variety of modulator and demodulator design options. Modulators can be designed from combinations of mixers, non-linear amplifiers, VCOs, diode circuits, and filters. Similarly, demodulators can be designed from combinations of electronic circuits. Many of the design options can be explored through discussion of amplitude modulation (AM) and frequency modulation (FM).

AM Modulation

In AM systems, it is typical to use mixers in the modulator to shift the baseband (or "raw" information such as audio) signal to a higher frequency to allow for transmission on an assigned carrier frequency. In AM radio receivers, the FCC approved radio frequency range is 540 kHz to 1600 kHz. Each radio station is allocated a frequency in that frequency range at 10 kHz increments.

In AM modulation, the baseband information alters the amplitude of a sinusoid at a significantly higher frequency than the highest baseband frequency. The high frequency sinusoid is called the carrier frequency. The heterodyning process is used to demodulate AM signals by shifting the AM radio frequency signal down to a lower *IF* frequency. AM radio receivers typically use 455 kHz *IF* electronics to process the modulated signal.

A simple Double-Side Band-Suppressed Carrier (DSB-SC) form of amplitude modulation using a mixer is shown in Figure 15.25.

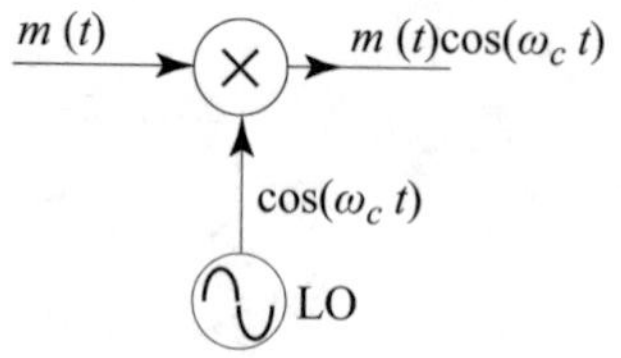

$m(t)$ $\longrightarrow$ $\times$ $\longrightarrow$ $m(t)\cos(\omega_c t)$

$\cos(\omega_c t)$

LO

Figure 15.25: DSB-SC mixer modulator.

In many instances, it is desirable to transmit not only the baseband information which has been up-converted to a higher frequency, but to also sent the carrier signal. Common examples of this type of AM are AM radio transmission where the station carrier frequency is transmitted with the up-converted information, and television signals. One possible method of transmitting a DSB plus a carrier (DSB+C) signal is shown in Figure 15.26.

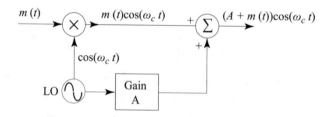

Figure 15.26: DSB+C modulator.

AM Demodulation

An envelope detector is the most common circuit used for converting of AM to baseband signal. An envelope detector is simply a lowpass filter allowing only the baseband signal to pass while eliminating the *IF* component generated after a mixer down-converts the modulated signal from the high carrier frequency. A simple envelope detector is shown in Figure 15.27. The diode allows only those signals greater than 0 V to pass through the detector. The input signal to the envelope detector is AM with carrier signal,

$$V_{IF}(t) = [A + m(t)]\cos\omega_{IF} t, \tag{15.56}$$

where

> A is a constant DC voltage
>
> $m(t)$ is the baseband signal
>
> ω_{IF} is the *IF* frequency (e.g., 455 kHz).

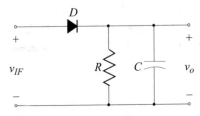

Figure 15.27: Envelope detection for AM.

An alternate technique to demodulate AM signals uses synchronous detection. The AM signal of Equation (15.56) is demodulated by multiplying the signal (using a mixer) by a *LO* signal of the same carrier frequency as shown in Figure 15.28, where $\omega_{IF} = \omega_c$.

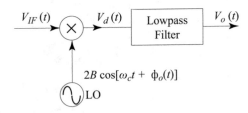

Figure 15.28: Synchronous direct conversion AM demodulator.

The output of the mixer for $\omega_{IF} = \omega_c$ is:

$$
\begin{aligned}
v_d\,(t) &= 2B\,[A + m\,(t)]\cos(\omega_{IF}t)\cos(\omega_{IF}t + \phi_o) \\
&= B\,[A + m\,(t)]\,[\cos(\omega_{IF}t - \omega_{IF}t - \phi_o) + \cos(\omega_{IF}t + \omega_{IF}t + \phi_o)] \quad\quad (15.57)\\
&= B\,[A + m\,(t)]\,[\cos\phi_o + \cos(2\omega_{IF}t + \phi_o)]\,.
\end{aligned}
$$

The mixer output consists of a baseband signal, $B[A + m(t)]\cos\phi_o$, and a modulated signal at the second harmonic frequency of the carrier, $B[A + m(t)]\cos(2\omega_{IF}t + \phi_o)$. The lowpass filter removes the high frequency component so that the resulting output is the demodulated baseband signal with some gain and *DC* bias,

$$
V_d\,(t) = B\,[A + m\,(t)]\cos\phi_o. \quad\quad (15.58)
$$

Unfortunately, since the input phase angle of the signal is not known, the output voltage can be very small. A phase difference $\phi_o = 0$ yields the maximum output voltage. This implies that the *LO* must be phase-locked to the carrier signal. In this case, the demodulation is truly one of coherent detection which has superior signal-to-noise performance over non-coherent detection methods such as with the use of envelope detectors. To insure that the *LO* is in phase-locked to the input carrier signal, a PLL can be used. The complete diagram of the PLL AM demodulator is shown in Figure 15.29. Since the phase detector causes the loop to lock with the VCO 90° out of phase with the input, a 90° phase shifter is placed between the output of the VCO and the mixer demodulator.

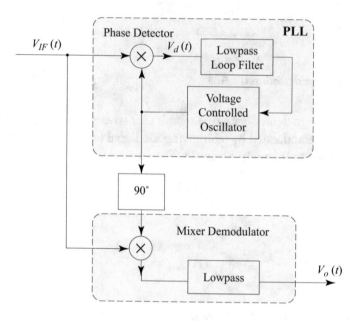

Figure 15.29: Phase-locked coherent AM demodulator.

RF frequency phase-shift networks that shift the phase by 90° at a single frequency are usually designed as *LC* circuits.

FM **Modulation**

In frequency modulated transmission, the baseband information alters the frequency of a sinusoid at a significantly higher frequency than the highest baseband frequency, whereas in phase modulated transmission, the baseband information alters the phase of a sinusoid. One of the most common methods used to generate frequency or phase modulated (*FM* and *PM*, respectively) signals is to use a VCO. By applying a time varying signal to reverse-bias a varactor diode in a VCO, an angle (generalized angle: either phase or frequency) modulated signal can be generated. One cannot distinguish the difference between an *FM* or *PM* signal by merely observing its time domain characteristics.

PM modulation is characterized by the following,

$$\varphi_{PM}(t) = A\left(\cos\omega_c t - k_p \sin\omega_c t\right), \tag{15.59}$$

and *FM* modulation is characterized by the following,

$$\varphi_{FM}(t) = A\left(\cos\omega_c t - k_f \sin\omega_c t \int m(\alpha)\,d\alpha\right),\tag{15.60}$$

where

φ is a voltage or current signal

k_f and k_p are *FM* and *PM* constants, respectively

A is a constant.

Other methods for generating *PM* and *FM* signals are shown in Figure 15.30, in which DSB-SC modulators are used for generation of the angle modulated signals.

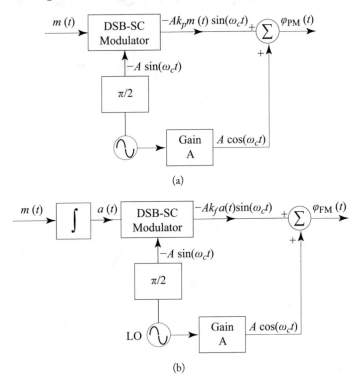

Figure 15.30: (a) *PM* modulator using DSB-SC modulator; (b) *FM* modulator using DSB-SC modulator.

FM Demodulation

There are a number of different methods for demodulating *FM* signals after down conversion to the *IF*. One common method uses a differentiator in series with an envelope detector. As shown in Figure 15.31. The output signal from an envelope detector is low pass filtered to provide an output signal proportional (by a constant k_f) to the original baseband signal.

Figure 15.31: *FM demodulation by direct differentiation.*

Another common type of *FM* demodulator is the discriminator circuit. This *FM* demodulator is often called a slope detector since the principle of detection is based on the detection of the signal amplitude variation as it traverses the slope of the resonant response of the circuit as shown in Figure 15.32.

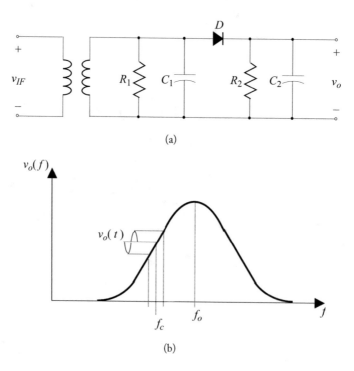

Figure 15.32: (a) Discriminator (slope detection) circuit; (b) Method of demodulating *FM* signals.

Although conceptually simple, discriminator circuits are sensitive to amplitude variations in the input signal. Variations in the input modulated signal will produce a falsely demodulated baseband signal. One solution is to apply the input modulated signal to a limiter prior to demodulation with a discriminator. However, this requirement has eliminated the use of discriminators in most production entertainment receiver circuits in favor of ratio detectors. In fact, although many integrated circuit receivers incorporate what are called discriminators, they are actually ratio detectors.

The ratio detector, shown in Figure 15.33, is similar in operation to discriminators except for its balanced peak-detector configuration which makes it less susceptible to amplitude variations in the input modulated signal.

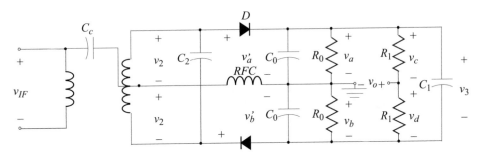

Figure 15.33: Ratio detector.

The ratio detector operates similarly to a half-wave rectifier with two diodes in series and an *RC* filter. The radio frequency choke (RFC) is placed in the circuit for *DC* isolation. The filtered output is v_3 which is proportional to $\eta|v_2|$ where v_2 is the *IF* phasor voltage on the secondary of the transformer, and η is the diode empirical scaling constant. $|v_2|$ and v_3 are essentially independent of frequency over the operating bandwidth of the detector. Since v_3 is divided across the two R_1 resistors and is symmetrical with respect to ground, v_c and v_d are of equal amplitude and

$$v_c = v_d = \frac{v_3}{2},\tag{15.61}$$

and

$$v_3 = v_c + v_d.\tag{15.62}$$

The phasor *IF* voltage, v_a', between the anode of the top diode and ground is,

$$v_a' = v_{IF} + v_2,\tag{15.63}$$

and the phasor *IF* voltage, v_b', between the cathode of the bottom diode and ground is,

$$v_b' = v_{IF} - v_2.\tag{15.64}$$

The rectified voltages across the two resistors of value R_0 are,

$$v_a = \eta\,|v_{IF} + v_2|,\tag{15.65}$$

and

$$v_b = \eta \, |v_{IF} - v_2| \, . \tag{15.66}$$

The output voltage across the resistor bridge and across the capacitor, C_1, are,

$$v_o = v_d - v_b, \tag{15.67}$$

and

$$v_3 = v_a + v_b. \tag{15.68}$$

Substituting Equations (15.61) and (15.68) into (15.67) yields the output voltage with respect to the voltages across the two resistors of value R_0,

$$v_o = \frac{v_3}{2} - v_b = \frac{v_a}{2} + \frac{v_b}{2} - v_b = \frac{v_a - v_b}{2}. \tag{15.69}$$

In steady-state, the diodes conduct during a small portion of each *IF* cycle. During conduction, the capacitors are recharged. The diodes will conduct if the instantaneous *IF* voltages v_a' and v_b' are greater than the instantaneous voltages v_a and v_b. The peak values of v_a' and v_b' drops and the diode conduction period cut short when the input signal level drops. For a given *FM* frequency deviation, $\Delta\omega$, the Q of the circuit on the secondary of the transformer increases making $\angle v_1 / v_2$ smaller. This increases the ratio detector output as $\Delta\omega$ increases.

A PLL can be used to demodulate *FM* signals. If the PLL is locked onto an input frequency, the control voltage for the VCO from the loop filter/amplifier is proportional to the VCO's shift in frequency from its free-running frequency. The control voltage shifts with a shifting input signal to the PLL. If the input to the PLL is an *FM* signal, the VCO control voltage from the loop filter/amplifier yields the demodulated output. The PLL is able to demodulate *FM* signals with a higher degree of linearity than other *FM* demodulation techniques.

15.7   RECEIVER DESIGN ISSUES

Many factors are used to determine rate receiver performance. The following specifications are commonly applied to a wide variety of communications receivers.

Sensitivity

Sensitivity is a measure of the weakest received signal that allows acceptable reproduction of the original signal. Ultimately, the sensitivity is limited by the noise generated by the receiver electronic circuits. Therefore, the receiver output noise is an important factor in quantifying sensitivity. Sensitivity is defined as the minimum carrier signal input voltage that will produce a specified signal-to-noise power ratio (SNR) at the output of the intermediate frequency (*IF*) section.

Noise Figure

The noise figure (*NF*) of a two-port network is a measure of the degradation of the SNR between the input and output terminals. A two-port network with noise is shown in Figure 15.34 with input signal power, P_{si}, and input noise power, P_{ni}, and corresponding output signal power, P_{so}, and output noise power, P_{no}.

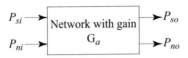

Figure 15.34: Two-port network with noisy input and output.

The noise figure as defined over a specified bandwidth is,

$$NF = \frac{\text{input SNR}}{\text{output SNR}} = \frac{P_{si}/P_{ni}}{P_{so}/P_{no}} = \frac{P_{no}}{G_a P_{ni}} = 1 + \frac{P_{ne}}{G_a P_{ni}}, \qquad (15.70)$$

where P_{ne} is the noise power generated by the two-port network.

The value of noise figure is most commonly expressed in terms of decibels:

$$NF_{dB} = 10 \log (NF), \qquad (15.71)$$

for a noise-free network, the input and output SNRs are equal and $NF = 1 = 0\,\text{dB}$.

Selectivity

Selectivity is a measure of the receiver's ability to capture a desired station and eliminate unwanted signals. This quality is determined in large part by the frequency response of the frequency selective circuits of the receiver.

Image Rejection

Image rejection is a measure of the attenuation of unwanted sum and difference frequencies produced by the mixer. Large values are desirable: typical values are about 50 dB.

Intermediate Frequency Rejection

The ratio of inputs at the *IF* and the desired carrier frequency that produce an equal output from the mixer is defined as the *IF* rejection ratio.

Intermodulation Distortion

Intermodulation distortion (IMD) is distortion products created by the non-linear response of the electronic circuit when excited by two (or more) sinusoidal inputs. IMD produces unwanted signals that may interfere with, and corrupt the desired signal.

The common theme that prevails in the receiver qualities mentioned are:

- Frequency selectivity

- Noise

- Distortion

By increasing frequency sensitivity, and decreasing noise and distortion in each subsystem.

15.8   CONCLUDING REMARKS

A sampling of communication electronic circuits was presented in this chapter. Many variants of the circuit discussed here are being used in a variety of communication system applications.

The circuits and concepts discussed in this chapter clearly shows that most of the circuits used can be designed and analyzed by the methods discussed thus far in this book. An analog-to-digital converter used OpAmp comparators to yield a digital representation of the time-varying input signal. A transistor mixer circuit is merely a small-signal amplifier with two inputs that take advantage of the nonlinear characteristics of the active device. A voltage controlled oscillator was shown to be an oscillator circuit with a voltage-variable reactive network. A phase-lock loop is a subsystem composed of a mixer, loop filter/amplifier, and a voltage controlled oscillator. The selection of filter configurations was shown to be dependent on the frequency range of operation, with passive filters used when the frequencies are higher than the audio range. Communications modulators and demodulators are designed using mixers, diode circuits, and PLLs.

Summary Design Example: Cable Television Converter

It is common to transmit the modulated information on a *RF* and microwave frequencies. Therefore, the received signal must be down-converted prior to demodulation. A down-converter is required by consumers with non-cable ready television sets in order to receive cable television signals. Cable television signals typically range in frequency from 55 MHz (channel 2) to 300 MHz to 1 GHz, depending on the number of standard television signals transmitted by the local cable television operator.[5] Since the cable television frequency assignments for channels greater than 13 are different from the broadcast channel frequency assignments, non-cable ready television sets will not be able to receive channels from the cable that are greater than 13.

To solve this problem, a converter unit is provided to the customer. The purpose of the converter unit is to down-convert all cable-transmitted channels to channel 3. The output of the converter unit is applied to the television set with its channel selector tuned to channel 3. Cable channel selection is performed by the converter unit. Other functions are built into the cable converter unit including de-scrambling capability to block reception of premium entertainment channels. The down-conversion to channel 3 (61.25 MHz carrier) is accomplished by a mixer. The allocated bandwidth per channel is 6 MHz. The frequency range allocated to channel 3 is 60 MHz–66 MHz.

[5]The frequency range depends on the number of standard NTSC format channels. Operational frequency range may be significantly different when High Definition Television (HDTV) signals are transmitted.

Several mixer topologies can be considered. The two mixer topologies considered for this design are:

- BJT active mixer

- Dual-gate FET mixer

For low cost, an inexpensive BJT active mixer may be most appropriate. Typical *RF* BJT specifications are:

$$\beta_F = 50, \ V_A = 150\,\text{V}, \ r_b = 30\,\Omega, \ f_T = 5\,\text{GHz at } 30\,\text{mA}$$
$$C_{ibo} = 8\,\text{pF at } V_{EB} = 0.5\,\text{V}, \ C_{obo} = 1\,\text{pF at } V_{CB} = 5\,\text{V}.$$

To down-convert channel 7 at a carrier frequency (*RF* frequency) of 174.25 MHz, the VCO-LO is set at $f_{LO} = 61.25\,\text{MHz} + 174.25\,\text{MHz} = 235.5\,\text{MHz}$, resistance $R_{LO} = 200\,\Omega$. The *IF* frequency is 61.26 MHz.

The mixer is to deliver 10 mW to a 50 Ω load with an efficiency of 50%. The output *RF* transformer has a primary inductance of 1.0 μH. A 12 V *DC* power supply is available.

The BJT mixer configuration is shown below:

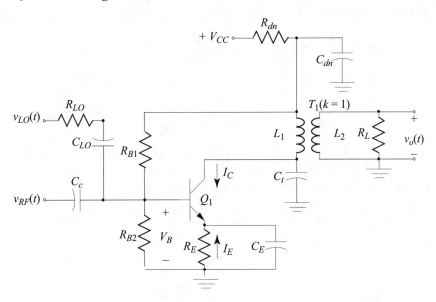

DC Design

For $V_E = 0.1V_{CC}$,

$$I_C = \frac{2P_O}{0.9V_{CC}} = \frac{2(0.01)}{0.9(12)} = 1.85\,\text{mA}.$$

For stable BJT operation,

$$R_E = -\frac{V_{CC}}{10 I_E} = \frac{V_{CC}(\beta_F + 1)}{10 \beta_F I_C} = \frac{12(51)}{10(50)(0.00185)} = 661 \approx 680\,\Omega,$$

$$R_{B2} = \frac{V_B}{10 I_B} = \frac{V_B \beta_F}{10 I_C} = \frac{\beta_F(-I_E R_E + V_{BE})}{10 I_C} = \frac{50[0.00189(680) + 0.7]}{0.0185} = 5.37\,\text{k} \approx 5.6\,\text{k}\Omega.$$

For quiescent point stability defined by a 1% (or less) change in collector current for a 10% change in β_F, the biasing rule-of-thumb is,

$$R_B \leq R_E \left(\frac{\beta_F}{9} - 1\right) = 680 \left(\frac{50}{9} - 1\right) = 3.1\,\text{k} \approx 3.0\,\text{k}\Omega,$$

where $R_B = R_{B1}//R_{B2}$.

Solving for R_{B1} yields, $R_{B1} \approx 6.2\,\text{k}\Omega$.

AC Design

Let the decoupling network resistance, $R_{dn} = 100\,\Omega$. The reactance of the capacitor C_{dn} must therefore be, $X_C \leq 10\,\Omega$, for 61.25 MHz. Therefore, the decoupling capacitor is, $C_{dn} \geq 260\,\text{pF}$ or $C_{dn} = 330\,\text{pF}$.

Since C_E is in series with C_{dn} in the AC path, let $C_E = C_{dn}$.

The emitter bypass capacitor C_E provides an AC low impedance for the transistor emitter,

$$X_{CE} = \frac{1}{10 g_m} = \frac{V_t}{10 I_C} = \frac{0.026}{10(0.00185)} = 1.4\,\Omega.$$

Therefore,

$$C_E = \frac{1}{\omega_{IF} X_{CE}} = \frac{1}{2\pi(61.25 \times 10^6)(1.4)} = 1850\,\text{pF} \approx 2200\,\text{pF}.$$

The value of the DC blocking capacitor C_C is determined in the same way as a bypass except that its reactance should be an order of magnitude less that the amplifier input impedance: that is, $X_{CC} = Z_{in}/10$, where Z_{in} is the amplifier input impedance.

$$X_{CC} = \frac{Z_{in}}{10} \approx \frac{R_B //(r_b + r_\pi)}{10} = \frac{3000//(30 + 703)}{10} = 59\,\Omega.$$

Therefore,

$$C_C = \frac{1}{\omega_{RF} X_{CC}} = \frac{1}{2\pi(174.25 \times 10^6)(59)} = 15.4\,\text{pF} \approx 15\,\text{pF}.$$

The coupling capacitor C_{LO} is,

$$C_{LO} = \frac{1}{2\pi f_{LO} R_{LO}} = \frac{1}{2\pi(235.5 \times 10^6)(200)} = 3.3\,\text{pF}.$$

The capacitor C_t is a tuning capacitance to tune to the proper Q and bandwidth at the *IF* frequency,

$$C_t = \frac{1}{(2\pi f_{IF})^2 L_1} = \frac{1}{\left[2\pi \left(61.25 \times 10^6\right)\right]^2 \left(1 \times 10^{-6}\right)} = 6.8 \, \text{pF}.$$

Since video signals are contained in a non-symmetric frequency range about the carrier frequency, with the carrier signal positioned at 1.25 MHz above the lower frequency corresponding to the lowest frequency of the channel range. The highest frequency corresponding to the channel range is 4.75 MHz above the carrier. Therefore, the design of the *IF* mixer bandwidth should contain $f_{IF} \pm 4.75 \, \text{MHz}$. In order to contain those frequencies, a bandwidth of $\pm 14.25 \, \text{MHz}$, which is wider than $\pm 4.75 \, \text{MHz}$, is used. The output from the mixer will be applied to a filter to appropriately shape the video signal.

Knowing the desired bandwidth of f_{IF} and BW_{IF}, the effective quality factor, Q_{eff} of the transformer is,

$$Q_{eff} = f_{IF}/BW_{eff} = 61.25 \times 10^6/14.25 \times 10^6 = 4.3.$$

Then the reactance of the primary of the transformer X_{L1} is,

$$R'_{L1} = X_{L1} \times Q_{eff} = 2\pi(61.25 \times 10^6)(10^6)(4.3) = 1.65 \, \text{k}\Omega.$$

The transformer turns ratio is,

$$n_p/n_s = \sqrt{R'_{L1}/R_L} = \sqrt{1650/50} = 5.7.$$

The input impedance of the filter which follows the mixer is $50 \, \Omega$.

15.9   PROBLEMS

15.1.   Design an 4-bit Flash ADC to digitize an analog signal with a peak-to-peak voltage of 5 V. Assume ideal OpAmps and powered by a $\pm 12 \, \text{V}$ power supply. Include the sample-and-hold circuit.

15.2.   A sample-and-hold circuit has a holding capacitor of 50 pF, and the leakage current in the HOLD mode is 1 nA. If the HOLD interval is 50 μs find the percentage output decay rate (called *droop*).

15.3.   A sample-and hold circuit has a holding capacitor of 100 pF, and the equivalent leakage resistance in the HOLD mode is 15 GΩ. Estimate the percentage output decay rate (droop) if the hold interval is 100 μs.

15.4.   In the circuit of Figure 15.4, let $R = 15 \, \text{k}\Omega$ and $C = 500 \, \text{pF}$. The input bias current of the output OpAmp is 300 nA. Estimate the percent output decay rate (droop) if the HOLD interval is 1 V.

15.5. Consider an 8-bit Flash ADC. If the voltage supply V_{CC} consists of a DC voltage V_+ and a ripple voltage with peak value $\pm\Delta$. Find Δ to insure that the error produced by Δ affects no bits other than the lowest significant bit (LSB).

15.6. Design a 1 MHz BJT-bias controlled Colpitts 1 MHz VCO with a ± 10 kHz tuning range, using a 3.3 μH inductor in the reactive feedback path. Find the frequency tuning range as a function of the bias control voltage, V_m, for the oscillator. The BJTs have identical parameters:

$$\beta_F = 200, \ V_A = 200 \, \text{V}, \ r_b = 30 \, \Omega, \ CJC = 14 \, \text{pF}, \ \text{and} \ f_T = 250 \, \text{MHz}.$$

15.7. Generate the graph of the MV2102 characteristics for reverse-bias voltages of 0 V to 20 V as shown in Figure 15.8. The varactor parameters are:

$$\Psi_0 = 0.7266 \, \text{V}, \ C_{jo} = 17.88 \, \text{pF}, \ \text{and} \ m = 0.424.$$

15.8. Complete the design of the VCO circuit shown, for a tuning range of $4.95 \, \text{MHz} \leq f_o \leq 5.05 \, \text{MHz}$. Determine the required range of the tuning voltage, V_m.

Assume that the transistor parameters are:

$$\beta_F = 200, \qquad\qquad V_A = 150 \, \text{V},$$
$$r_b = 30 \, \Omega, \qquad\qquad f_T = 250 \, \text{MHz},$$
$$C_{ibo} = 6.5 \, \text{pF} \quad \text{at} \quad V_{EB} = 0.5 \, \text{V},$$
$$C_{obo} = 3.3 \, \text{pF} \quad \text{at} \quad V_{CB} = 5 \, \text{V}.$$

Simulate the circuit using SPICE and confirm the oscillation frequency.

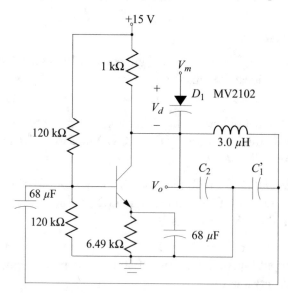

15.9. Design an NMOSFET-based Colpitts VCO at with a tuning rage of $1.97\,\text{MHz} \leq f_o \leq 2.03\,\text{MHz}$ using an MV2102 varactor diode. Determine the required range of the tuning voltage, V_m. The FET parameters of interest are:

$$I_{DSS} = 6\,\text{mA}, \quad V_{PO} = -4.7\,\text{V}, \quad V_A = 100\,\text{V},$$
$$C_{iss} = 4.5\,\text{pF at } V_{GS} = 0\,\text{V}, \text{ and } C_{rss} = 1.5\,\text{pF at } V_{GS} = 0\,\text{V}.$$

A + 15 V power supply is available.

Simulate the circuit using SPICE and confirm the oscillation frequency. How is the oscillation frequency affected when the temperature is changed to 50°C?

15.10. Design a 1 MHz Hartley VCO with a ± 10 kHz tuning range, using an auto-transformer and an MV2102 varactor diode. Find the frequency tuning range as a function of the diode tuning voltage, V_m. The BJT parameters are:

$$\beta_F = 200, \quad V_A = 200\,\text{V}, \quad r_b = 30\,\Omega, \quad CJC = 14\,\text{pF}, \text{ and } f_T = 250\,\text{MHz}.$$

15.11. Design an active mixer using a BJT with the following parameters:

$$\beta_F = 200, \quad V_A = 150\,\text{V}, \quad r_b = 30\,\Omega,$$
$$C_{ibo} = 3\,\text{pF at } V_{EB} = 0.5\,\text{V}, \quad C_{obo} = 1\,\text{pF at } V_{CB} = 5\,\text{V}, \text{ and } f_T = 750\,\text{MHz}.$$

The mixer is to deliver 6 mW to a 50 Ω load with an efficiency of 40%. The *RF* input frequency is 100 MHz and the *IF* is 10.7 MHz. The primary inductance of the output transformer is 0.3 mH (infinite Q). The *IF* bandwidth is 500 kHz. Use SPICE to confirm the operation of the mixer (inspect the frequency spectra).

15.12. Design an active mixer using a BJT with the following parameters:

$$\beta_F = 200, \quad V_A = 150\,\text{V}, \quad r_b = 30\,\Omega,$$
$$C_{ibo} = 7\,\text{pF at } V_{EB} = 0.5\,\text{V}, \quad C_{obo} = 4\,\text{pF at } V_{CB} = 5\,\text{V}, \text{ and } f_T = 350\,\text{MHz}.$$

The mixer is to deliver 15 mW to a 75 Ω load with an efficiency of 40%. The *RF* input frequency is 2 MHz and the *IF* is 455 kHz. The primary inductance of the output transformer is 0.3 mH (infinite Q). The *IF* bandwidth is 25 kHz. Use SPICE to confirm the operation of the mixer (inspect the frequency spectra).

15.13. For the dual-gate FET mixer shown, determine:

   (a) the bias voltage at both gates

   (b) the drain and source bias voltages for $I_D = 2\,\text{mA}$.

(c) the value of L in order to use a 10 pF (max) variable capacitor at mid capacity if the stray and FET drain capacitances total 3 pF for an *IF* of 45 MHz.

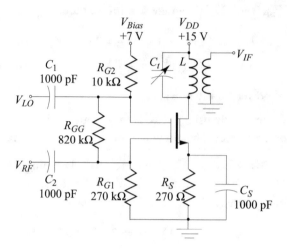

15.14. Define *phase-locked*.

15.15. A phase comparator with triangular transfer characteristic has a maximum output voltage of 4 V. Determine the gain in Volts/radian and Volts/degree of phase.

15.16. An analog phase detector with two input generators (no VCO) has a beat-frequency output of 4 V peak-to-peak at 100 Hz. Determine the phase detector gain (sensitivity) in Volts/radian and Volts/degree of phase.

15.17. A VCO is linear between 250 kHz and 330 kHz. The corresponding input voltages are 220 mV and − 220 mV, respectively.

    (a) Determine the VCO gain (sensitivity).

    (b) Determine the free-running frequency.

15.18. For a PLL with $K_d = 0.5$ V/rad, $K_a = -4$, and $K_o = 30$ kHz/V, a VCO free-running frequency of 200 kHz and a triangular characteristic:

    (a) Determine the PLL loop gain.

    (b) Determine the VCO input voltage for the PLL locked to a 180 kHz input signal.

    (c) Determine the maximum voltage output from the phase detector.

    (d) Determine the hold-in frequency range.

15.19. Complete the system design of the PLL shown for a loop gain of 100 dB at 1 rad/s. Determine the static phase error, ϕ_e, and V_d when the PLL is phase-locked.

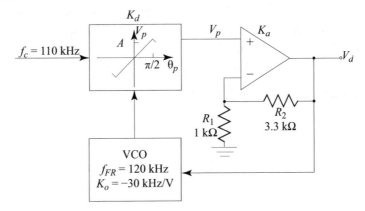

15.20. Design a capacitor coupled constant-k bandpass filter centered at 10.7 MHz $\pm$ 200 kHz, 1 dB maximum ripple, 800 kHz $\pm$ 100 kHz 3 dB bandwidth, and 25 dB attenuation at 11.5 MHz. The source and load resistors are 75 Ω and the available inductors have a quality factor of 70. Confirm the design using SPICE.

15.21. A high-selectivity AM receiver has a 455 kHz *IF* with a 10 kHz, 3 dB bandwidth requirement. Interference at 427.5 kHz coming through the mixer and *RF* amplifier must be reduced by 48 dB. Assume a 50 Ω system. The available inductors have a $Q_u = 100$. Design the filter and confirm its operation using SPICE.

15.22. Perform a SPICE simulation of Example 15.4.

15.23. Design a capacitor coupled bandpass filter with a minimum number of resonators, ripple ≤ 0.3 dB, center frequency = 2 MHz, 3 dB bandwidth = 100 kHz, and 36 dB attenuation at 2169 kHz, in a 1 kΩ system. The available inductors have a $Q_u = 100$. Confirm the design using SPICE.

15.24. A 2.11 GHz, 28.6 MHz bandwidth satellite receiver must reject an adjacent channel transmitter by 60 dB. Carriers are 36 MHz apart. The filter passband ripple must not exceed 0.5 dB, and the insertion loss must not exceed 4.4 dB. A cavity filter with $Q_u = 1000$ is used.

    (a) How many resonators are required?

    (b) What is the actual ripple?

    (c) What will be the insertion loss?

15.25. Design a single-tuned FET amplifier with a center frequency at 500 kHz, $Q = 50$, and gain $|A_v| > 5$ using an inductor $L = 10 \,\mu$H with $Q_u = 200$. The source and load resistors are each 1 kΩ. The FET parameters are: $I_{DSS} = 5$ mA, $V_{PO} = -2$ V, $V_A = 150$ V, $C_{rss} = 6.5$ pF, $C_{iss} = 35$ pF. Confirm the design using SPICE.

15.26. Design a single-tuned BJT amplifier with a center frequency at $220\,\text{kHz}$, $Q = 50$, and gain $|A_v| > 5$ using an inductor $L = 10\,\mu\text{H}$ with $Q_u = 200$. The source and load resistors are each $1\,\text{k}\Omega$. The BJT parameters are: $\beta_F = 200$, $f_T = 400\,\text{MHz}$, $V_A = 200\,\text{V}$, and $C_{obo} = 2\,\text{pF}$. Confirm the design using SPICE.

15.27. Complete the design of the envelope detector shown for an *IF* frequency of $455\,\text{kHz}$, $V_\gamma = 0.7\,\text{V}$, and $P = 1\,\text{k}\Omega$.

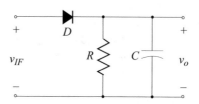

15.28. Design a LC $\pi/2$ phase-shifter at an *IF* frequency of $10.7\,\text{MHz}$. What is the signal attenuation at the *IF*?

15.29. Design a direct differentiation *FM* demodulator for an *IF* frequency of $1\,\text{MHz}$.

15.30. Derive the total system noise figure, NF_{sys}, for the two amplifier system shown below. P_{na1} and P_{na2} are the noise powers generated by the two amplifiers. The amplifier gains are G_{a1} and G_{a2}, with corresponding noise figures N_{F1} and N_{F2}.

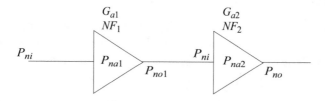

REFERENCES

[1] Ghausi, M. S., *Electronic Devices and Circuits: Discrete and Integrated*, Holt, Rinehart and Winston, New York, 1985.

[2] Hayward, W. H., *Introduction to Radio Frequency Design*, Prentice-Hall, Englewood Cliffs, 1982.

[3] Krauss, H. L., Bostian, C. W., and Raab, F. H., *Solid State Radio Engineering*, John Wiley & Sons, New York, 1980.

[4] Lathi, B. P., *Modern Digital and Analog Communication Systems, 2nd Ed.*, Holt, Rinehart, and Winston, Philadelphia, 1989.

[5] Millman, J. and Halkias, C. C., *Integrated Electronics: Analog and Digital Circuits and Systems*, McGraw-Hill Book Company, New York, 1972.

[6] Roden, M. S., *Analog and Digital Communication Systems*, *3rd Ed.*, Prentice-Hall, Englewood Cliffs, 1991.

[7] Sedra, A. S. and Smith, K. C., *Microelectronic Circuits*, *3rd Ed.*, Holt, Rinehart, and Winston, Philadelphia, 1991.

[8] Schilling, D. L. and Belove, C., *Electronic Circuits*, *3rd Ed.*, McGraw-Hill Book Company, New York, 1989.

[9] Smith, J., *Modern Communication Circuits*, McGraw-Hill Book Company, New York, 1986.

[10] Young, P. H., *Electronic Communication Techniques*, *3rd Ed.*, Merrill Publishing Company, New York, 1994.

CHAPTER 16

Digital Circuits

A digital electronic circuit is a device that operates on single or multiple input signals to produce an output that is limited to one of a few possibilities. The most common circuits are binary digital circuits: those circuits that have a single output limited to only two output states. The two-state behavior of digital circuits leads to referring to two-state a circuit as a "gate": it is either open or closed—ON or OFF. Gates are often connected in series with other gates. Gates that drive others are "master" gates: gates that are driven are "slave" gates. A single gate can perform both master and slave operations to individual surrounding gates.

Short introductions to the operation of selected binary digital circuits are presented in the introductory chapters on transistor functionality. Chapter 3 (Book 1) discusses the essential operating principles of two bipolar logic families using simple linear models of the BJT. Chapter 4 (Book 1) discusses MOS logic inverters using the principles of load lines. While it is assumed that the those chapters are prerequisites to this chapter, several digital circuit operating principles are presented that warrant repetition. Among the most important are:

- Logic Voltage Levels,

- Noise Margin, and

- Fan-out.

The output of a digital circuit is characterized by two voltage levels: a logic HIGH voltage and a logic LOW voltage. These voltages are symbolized as V_{oH} and V_{oL}, respectively. The input is also characterized by two voltage levels: the level above which all inputs are a logic HIGH, $V_{iH(\min)}$, and the level below which all inputs are a logic LOW, $V_{iL(\max)}$. For purposes of noise immunity it is important that the output and input voltage levels not be the same. Specifically

$$V_{oH} > V_{iH(\min)} \quad \text{and} \quad V_{oL} < V_{iL(\max)}.$$

The measures of this noise immunity are the *noise margins* NM(HIGH) and NM(LOW):

$$\text{NM(HIGH)} = V_{oH} - V_{iH(\min)},$$

and

$$\text{NM(LOW)} = V_{iL(\max)} - V_{oL}.$$

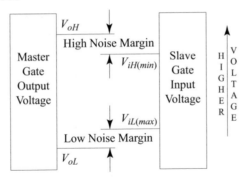

Figure 16.1: A physical interpretation of noise margin.

A descriptive diagram of the noise margins as well as slave input and master output voltage levels is shown in Figure 16.1.

Fan-out is a measure of the number of similar slave gates that a master gate can drive without producing logical errors. Typically, current loading determines the fan-out of a gate, but, as in the case of MOS gates, gate speed can be the determining factor. Fan-in is another term often found in the literature: it identifies the number of gate inputs.

While it is impossible to completely describe all types of digital circuits and all aspects of their operation in a single chapter, this chapter strives to present the *essential aspects* of the major digital circuits commonly in use. The fundamentals of the speed of digital logic transitions begins the discussion of both bipolar and MOS gates. Three families of bipolar gates and two families of MOS gates are discussed extensively. Regenerative logic circuits (latches, flip-flops, and Schmitt triggers) arepresented as are the fundamentals of memory circuits. The chapter ends with a descriptive section on Gallium Arsenide logic circuits.

16.1   THE SWITCHING SPEED OF BIPOLAR JUNCTION TRANSISTORS

Bipolar Digital Logic gates depend on the transition of the output of a BJT from one logic level to another. In the case of classic TTL logic gates this transition takes place between the cut-off and saturation regions of the output transistor. The speed at which any gate operates is limited by the transition speeds of its constituent transistors. Similarly, the maximum clock rate at which a synchronous digital system will operate is limited by transistor transition speed. The transition speed of a transistor is a function of the physical parameters of the transistor and of the components in the circuit surrounding the transistor. While it is impossible to analyze the effect of all possible circuit topologies on switching speed, an analysis of the switching speed the simple BJT inverter provides much insight into all BJT switching.

As shown in Figure 16.2, a general BJT inverter consists of a common-emitter BJT with Thévenin sources connected to the base and collector terminals. These Thévenin sources represent the surrounding logic gate circuitry.

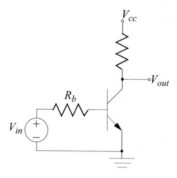

Figure 16.2: A simple BJT inverter.

As the input voltage changes between logic levels, the output will change to the opposite logic levels. This transition can not take place instantaneously: various delays must occur. Conceptually, the response of simple BJT inverter to a rectangular pulse is shown in Figure 16.3.

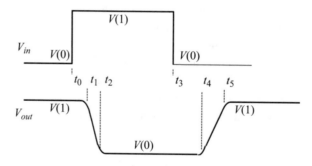

Figure 16.3: The response of an inverter to a digital pulse.

The pulse response consists regions of constancy and regions of transition. In the regions of transition there are four significant time periods.

- Delay Time $= t_d = t_1 - t_0$

- Rise Time $= t_r = t_2 - t_1$

- Storage Time $= t_s = t_4 - t_3$

- Fall Time $= t_f = t_5 - t_4$

The delay time is the time between the pulse transition and when the response transitions 10% of the distance between HIGH and LOW states. The rise time is the time for a 10% to 90% transition from HIGH to LOW. The term *rise time* refers to the BJT collector current change: as the voltage transitions from a HIGH to LOW, the collector current rises from a minimum to a maximum value. The storage time and fall time measure the equivalent time periods in the LOW to HIGH transition.

Rise Time

Perhaps the easiest region to analyze is the rise time. Here the BJT is in the forward-active region of operation. In this region of operation, the transistor speed is most often described by the forward time constant, τ_F, or by its frequency-domain equivalent, the unity-gain frequency, ω_T.[1]

$$\tau_F \approx \frac{1}{\omega_T}. \tag{16.1}$$

While the forward time constant is a useful parameter (for example, it is necessary in SPICE analysis), a more relevant parameter for gate speed calculations is the forward-active region time constant, τ_f:

$$\tau_f = \beta_F \tau_F \approx \frac{\beta_F}{\omega_T} = \frac{1}{\omega_{3\,\mathrm{dB}}}. \tag{16.2}$$

Transitions between two steady-state levels, X_i and X_f, in the forward-active region are described by a simple exponential relationship with this time constant:

$$X(t) = X_f - \left[X_f - X_i \right] e^{\frac{-(t-t_0)}{\tau_f}}. \tag{16.3}$$

It can easily be shown that rise time of a simple exponential transition is:

$$\text{rise time} \approx 2.2\tau_f. \tag{16.4}$$

Unfortunately, a transition entirely in the forward-active region is the type of transition typical of BJT logic gates. In a BJT gate, the LOW logic level is characterized by the saturation region of the BJT. Thus, the apparent final steady-state level is not the same as the actual final level. The BJT collector rise is consistent with a final collector current, $I_{cf} = \beta_F I_b$ and the time constant τ_f. The transition is completed when the collector current reaches its saturation value, $I_{c(sat)} < \beta_F I_b$. A significantly reduced rise time results. Figure 16.4 is a graphical interpretation of the reduction in rise time.

The rise time can be calculated by determining the 90% and 10% times, t_2 and t_1, respectively. Assume the external circuit parameters define the two currents, $I_{c(sat)}$ and I_{b1}:

$$I_{c(sat)} = \frac{V_{cc} - V_{ce(sat)}}{R_c}, \tag{16.5}$$

---

[1]The unity gain frequency is defined as the radian frequency at which the common-emitter gain is unity. It is discussed fully in Section 10.4 (Book 3).

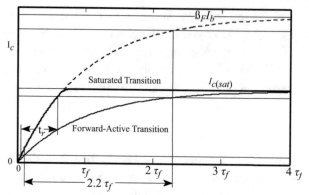

Figure 16.4: Saturated transition rise time.

and

$$I_{b1} = \frac{V_i - V_{be}}{R_b}. \tag{16.6}$$

For a saturated LOW output, define the current ratio, N_1, as:

$$N_1 = \frac{\beta_F I_{b1}}{I_{c(sat)}}. \tag{16.7}$$

N_1 is an indicator of the degree in which the BJT has been driven into the saturation region: it is called the *saturation overdrive factor*. Values of N_1 near unity indicate that the BJT is barely into the saturation region: larger values indicate a large excess of base current over what is necessary to saturate the BJT. The time for the collector current to achieve 90% of the saturated collector current is the solution to the expressions:

$$0.9 I_{c(sat)} = N_1 I_{c(sat)} \left(1 - e^{\frac{-t_2}{\tau_f}}\right). \tag{16.8}$$

Similarly, the time to achieve 10% of the saturated collector current is the solution to:

$$0.1 I_{c(sat)} = N_1 I_{c(sat)} \left(1 - e^{\frac{-t_1}{\tau_f}}\right). \tag{16.9}$$

The rise time, t_r, is the difference of t_2 and t_1:

$$t_r = t_2 - t_1 = \tau_f \ln\left(\frac{N_1 - 0.1}{N_1 - 0.9}\right). \tag{16.10}$$

As N_1 increases the rise time is made smaller. This seems to indicate that a strongly saturated BJT is desirable in terms of gate speed: other factors will show strong saturation is not desirable.

Delay Time

The delay time indicates the time between the input signal transition and the start of the rise time. Its primary components[2] are:

- t_{d1}—the time for minority carriers to transit the base and reach the collector

- t_{d2}—the time for the collector current to rise to 10% of its final value

Statistical analysis shows that the first of these factors is given by:

$$t_{d1} \approx \frac{\tau_F}{3} = \frac{\tau_f}{3\beta_F}. \tag{16.11}$$

The second factor is a portion of the exponential collector current change and can be derived in the same fashion. The 0% to 10% portion of the collector current rise takes place in time:

$$t_{d2} = \tau_f \ln\left(\frac{N_1}{N_1 - 0.1}\right). \tag{16.12}$$

The delay time is the sum of these factors.

$$t_d = t_{d1} + t_{d2} \approx \tau_f \left[\ln\left(\frac{N_1}{N_1 - 0.1}\right) + \frac{1}{3\beta_F}\right]. \tag{16.13}$$

Fall Time

The fall time, t_f, is analogous to the rise time. It indicates the time for the BJT collector current to fall from 90% to 10% of its saturated value. This fall in collector current occurs as the transistor transition in the forward-active region from saturation to cut-off. The expression for fall time is therefore:

$$t_f = t_4 - t_3 = \tau_f \ln\left(\frac{N_2 - 0.9}{N_2 - 0.1}\right). \tag{16.14}$$

Here a reverse overdrive factor, N_2, is defined as the ratio of β times the instantaneous base turn-off current to the saturation collector current:

$$N_2 = \frac{\beta_F I_{b2}}{I_{c(sat)}}, \tag{16.15}$$

where

$$I_{b2} = \frac{-V_{be(\text{active})}}{R_b}. \tag{16.16}$$

Notice that N_2 is a *negative* quantity. Large magnitude N_2 indicates a short fall time.

---

[2]Another factor is the time required to charge the base-emitter junction to the cut-in voltage. Here it is assumed that this factor is small compared to the other delay time factors.

Storage Time

The storage time indicates the time between the input signal transition and the start of the fall time. Its primary components[3] are:

- t_{s1}—the time for minority carriers to transit the base and reach the collector

- t_{s2}—the time for the collector current to rise to 10% of its final value

- t_{s3}—the time to dissipate the excess charge stored in the base of the saturated BJT.

The first two factors are direct analogs of similar components in the delay time:

$$t_{s1} \approx \frac{\tau_F}{3} = \frac{\tau_f}{3\beta_F},$$

(16.17)

and

$$t_{s2} = \tau_f \ln\left(\frac{N_2 - 1}{N_2 - 0.9}\right).$$

(16.18)

The third factor, t_{s3}, is related to an exponential decay of base charge when the BJT is in the saturation region. This decay has a time constant, τ_s, that is a function of the forward-active region time constant, τ_f, and its inverse-active region counterpart, τ_r. In most situations these two time constants are approximately equal and

$$\tau_s \approx \tau_f + \beta_R(\tau_f + \tau_r) \approx \tau_f + \beta_R(\tau_f + \tau_f) = \tau_f(1 + 2\beta_R).$$

(16.19)

This dual dependence is due to the forward biased condition of both the base-emitter and the base-collector junctions of a saturated BJT. As a consequence, the saturation time constant is significantly longer than either of the other time constants. The time for base charge dissipation is given by:

$$t_{s3} = \tau_s \ln\left(\frac{N_1 - N_2}{1 - N_2}\right).$$

(16.20)

The total storage time is given by the sum of the individual components:

$$t_s \approx \frac{\tau_f}{3\beta_F} + \tau_f \ln\left(\frac{N_2 - 1}{N_2 - 0.9}\right) + \tau_s \ln\left(\frac{N_1 - N_2}{1 - N_2}\right).$$

(16.21)

Large magnitude N_2 (a negative quantity) will decrease the storage time. However, large N_1 will significantly *increase* the storage time. This is especially significant since the storage time constant is larger than the forward-active region time constant, $\tau_s > \tau_f$. If the transistor does not enter the saturation region ($N_1 = 1$), $t_{s3} = 0$ and the storage time is composed of only the first two components of Equation (16.21).

---

[3]Another factor is the time required to charge the base-emitter junction to the cut-in voltage. Here it is assumed that this factor is small compared to the other delay time factors.

Summary

While the switching speed of BJT is largely dependent on the physical parameters of the transistor itself, the surrounding circuit parameters also have significant effect. Rise time and fall time are strongly dependent on the overdrive factors, N_1 and N_2, respectively. In each case, an increase in the magnitude of the overdrive factor reduces the respective time. Delay time is also decreased by large N_1 as is storage time by large magnitude N_2. Unfortunately storage time can be greatly *increased* by large N_1 (saturated BJTs). The propagation delay of a TTL inverter can be described in terms of the transition times derived. Its component terms are:

$$t_{PHL} \approx t_d + \frac{t_r}{2} \quad \text{and} \quad t_{PLH} \approx t_s + \frac{t_f}{2}. \tag{16.22}$$

The average propagation delay is given by:

$$t_{PD} = \frac{t_{PHL} + t_{PLH}}{2} \approx \frac{1}{2}\left(t_d + t_s\right) + \frac{1}{4}\left(t_r + t_f\right). \tag{16.23}$$

16.2   BIPOLAR DIGITAL GATE CIRCUITS

Bipolar logic gates are divided into four fundamental logic gate families: diode-transistor logic (DTL), transistor-transistor logic (TTL), emitter-coupled logic (ECL), and integrated injection logic (I^2L). While the original form of DTL is now obsolete, these four families of gates form the foundation for all modern bipolar logic gates. Advancements in the design of these gates that have taken place are generally for the purpose of improving gate performance in one or more of the following areas:

- Speed of operation

- Power consumption

- Noise rejection

- Fan-out

Unfortunately, an improvement in one aspect of gate performance may degrade performance in another. Good design is a balance of often-conflicting design constraints: modern logic gate design is an example of the balancing of these constraints. It is the purpose of this section to explore some of the designs common in digital gates.

The principles of operation of several basic digital logic gates that use bipolar junction transistors is presented, using quasistatic analysis techniques, in Section 3.5 (Book 1). Power consumption, fan-out, and many aspects of noise rejection can be analyzed using these simple, linear techniques. Determination of the speed at which a bipolar logic gate operates can be using the basic techniques presented in Section 16.1. Since TTL found its origins in DTL, modern design in the two logic families will be discussed together. ECL and I^2L will be discussed separately.

16.2.1 TTL AND DTL LOGIC GATES

The output voltage levels of a common DTL and TTL logic circuit is developed as the output BJT switches between two regions of operation: cut-off and saturation. It is these two stable states that make these families of gates reliable and predictable.

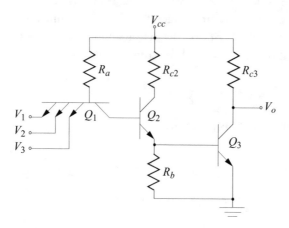

Figure 16.5: Basic TTL NAND gate.

The NAND gate forms the fundamental logical unit in both DTL and TTL logic: all other gate logic operations can be derived from this fundamental unit. The basic circuit topology of a simple TTL NAND logic gate is shown in Figure 16.5. This gate is typically operated with a supply voltage of 5 V and consists of several resistors and three BJTs. The input BJT, Q_1, is constructed with multiple emitters (three emitters are shown: other numbers are common) which serve as the individual logic inputs to the gate. When any of the logic inputs is a logic LOW, Q_1 enters the saturation region. There is an insufficiently high voltage at the base of Q_2 to forward bias the base-emitter junctions of Q_2 and Q_3, thus Q_2 and Q_3 are OFF and a logic HIGH output is produced. A LOW logic output occurs when all inputs are HIGH. When this input condition occurs, Q_1 enters the inverse-active region.[4] Q_2 turns ON and forces Q_3 into the saturation region, production a logic LOW. Depending on exact resistor values and BJT parameters, Q_2 typically enters the saturation region, although not as strongly as Q_3. The nominal output logic levels for this simple, unloaded gate are:[5]

$$V_{oH} = 5\,\text{V} \quad \text{and} \quad V_{oL} = 0.2\,\text{V}.$$

[4]In many situations (large fan-out), the master gate can not supply enough current to sustain the inverse-active region of Q_1 in the slave gate. In that case, Q_1 is in an *inverse* saturation state: both junction are forward biased, but the base-collector junction is *more* forward biased than the base-emitter junction.
[5]Logic levels, noise margins, and fan-out for this gate are calculated in Section 3.5 (Book 1).

The extremes of the input levels are:

$$V_{iH(\min)} = 1.8\,\text{V} \quad \text{and} \quad V_{iL(\max)} = 1.1\,\text{V}.$$

These levels result in noise margins of:

$$\text{NM(HIGH)} = V_{oH} - V_{iH(\min)} = 5 - 1.8 \approx 3.2\,\text{V}.$$
$$\text{NM(LOW)} = V_{iL(\max)} - V_{oL} = 1.1 - 0.2 \approx 0.9\,\text{V}.$$

While this simple gate topology operates well, it has several properties that are undesirable for current IC realizations. Most problematic are its relatively high power consumption and low speed of operation.

The low speed operation of simple TTL and DTL logic gates stems from junction charge buildup in BJTs that enter the saturation region. As was seen with the simple bipolar inverter, strongly saturated BJTs exhibit a relatively long storage time. This long storage time slows the digital transitions necessary in a logic gate. In the TTL NAND gate under analysis, both the logic LOW and logic HIGH states have BJTs in saturation. A logic LOW implies that Q_1 is in saturation and a logic HIGH, implies that Q_2 and Q_3 are in saturation.

Each junction must dissipate its stored charge when the gate transitions its output between logic levels. The speed of the charge dissipation strongly depends upon the Thévenin resistance apparent to the junction. It is in this Thévenin resistance that a conflict between two TTL gate design goals is most apparent. Any attempt to increase the speed of the gate by reducing the resistance values results in an increase in the gate average power consumption. Similarly, reducing the average power consumption by increasing the resistor values results in slower gate performance. With this basic gate topology, power consumption and gate speed can only be simultaneously improved by lowering the supply voltage: unfortunately this action reduces the HIGH noise margin, NM(HIGH).

Gate performance can be improved in several areas simultaneously only with changes in the basic topology of the gate. Historically these changes have taken place gradually and have resulted in a series of TTL gate families. The changes have been centered on two basic design techniques:

- active charge dissipation, and

- transistor saturation control.

While it is not feasible to extensively discuss each variation of gate topology in this text, a brief look at several TTL gate alterations is instructive in the study of logic gate speed.

Active Charge Dissipation

In typical TTL gates the output BJT, Q_3, is the most strongly saturated transistor. In addition, the active elements surrounding Q_1 and Q_2 provided relatively low-impedance paths for rapid dissipation of built-up charge. Therefore, the greatest benefit is obtained by focusing efforts at increasing the gate speed about the output transistor. For a LOW output, Q_3 is saturated: both base-emitter and base-collector junctions are forward biased. While the charge build-up is largest in the base-emitter junction, it is the base-collector junction that must undergo the greatest change in charge distribution as Q_3 switches state from saturation to cut-off. In order to quickly remove the charge from the base-collector junction, the collector resistor is replaced by a low-impedance active load, as shown in Figure 16.6. This particular circuit topology is referred to as an active pull-up or sometimes a "totem-pole" output.

With active pull-up output, the saturated transistor, Q_3, sees a Thévenin load at the collector consisting of the dynamic resistance of a diode in series with the output resistance of a common-collector amplifier:

$$R_{th} \approx r_d + \frac{R_{c2}}{\beta_{F4} + 1}.$$

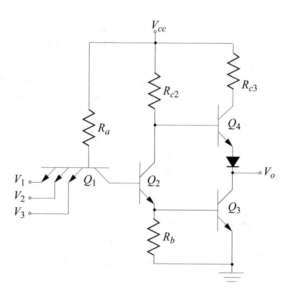

Figure 16.6: TTL gate with active pull-up.

This apparent resistance is much smaller than that of the collector resistor in the simple TTL topology while the Thévenin voltage has not significantly changed. These changes have the effect of reducing the overdrive factor, N_1, during discharge:

$$N_{1(\text{discharge})} = \frac{\beta_F I_b}{I_{c(\text{eff})}},$$

where

$$I_{c(eff)} = \frac{\left(V_{cc} - V_{be(active)} - V_\gamma\right) - V_{ce(sat)}}{R_{th}} > I_{c(sat)}.$$

The storage time for Q_3 is significantly reduced thereby reducing the collector current fall time and increasing the gate speed. Since the actual value of the collector current at saturation, $I_{c(sat)}$, remains essentially unchanged, active pull-up has little effect on the collector current rise time.

Gate operation with active pull-up is much the same as with a resistor as the output collector load. The input circuitry of the TTL gate remains unchanged, therefore the input logic voltage levels remain unchanged. The output LOW voltage remains the same at $V_{CE(sat)} \approx 0.2$ V: only the output HIGH voltage changes. For a logic HIGH output, Q_2 and Q_3 are cut-off. It is assumed that a load attached to the output of the gate draws some current, thus Q_4 is in the forward-active region. The nominal output voltage level is the source voltage reduced by the voltage drops across the resistor, R_{c2}, the base-emitter junction of Q_4, and the diode:

$$V_{oH} = V_{cc} - \left(V_{R_{c2}} + V_{BE4(on)} + V_\gamma\right) \approx V_{cc} - 1.5 \approx 3.5 \text{ V}.$$

The reduction in the nominal output HIGH level decreases the HIGH noise margin, NM(HIGH):

$$\text{NM(HIGH)} = V_{oH} - V_{iH(min)} = 3.5 - 1.8 \approx 1.7 \text{ V}.$$

This form of active pull-up is the output stage found in 74XX/54XX series TTL gates. Added benefits of this active pull-up circuit topology are a decrease in average power consumption of 10–20% over the simple resistive pull-up configuration and a very slight improvement in fanout. Both improvements are due to Q_4 being in the cut-off region of operation for a LOW output.

Example 16.1

Determine the fan-out of the active pull-up NAND gate shown in Figure 16.6. The pertinent circuit parameters are:

$$V_{cc} = 5 \text{ V} \qquad R_a = 3.9 \text{ k}\Omega \qquad R_b = 1.0 \text{ k}\Omega$$
$$R_{c2} = 1.5 \text{ k}\Omega \qquad R_{c3} = 130 \,\Omega.$$

The physical realization is in Silicon with BJT parameters:

$$\beta_F = 50 \qquad\qquad \beta_R = 2.$$

Solution:

The determination of fan-out is much the same as that of the resistive pull-up circuit (Example 3.5 (Book 1)). Fan-out in TTL gates is determined by a master gate with a LOW output driving a slave gate. The input current for a slave gate with a low input (0.2 V) is found to be:

$$I_{in} = \frac{5.0 - 0.2 - 0.8}{3.9 \text{ k}} = 1.026 \text{ mA}.$$

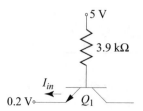

If the master gate is driven by other gates of the same type, it is unreasonable to assume that large amount of current is entering the emitter of Q_1 (it would draw the input voltage below zero). For fan-out calculations it is safer to assume the worst case scenario where the input current to the master gate is approximately zero. Under that scenario

$$I_{B2} \gg I_{B1}$$

$$I_{B2} \approx \frac{5 - (0.8 + 0.8 + 0.7)}{3.9\,k} = 692\ \mu A.$$

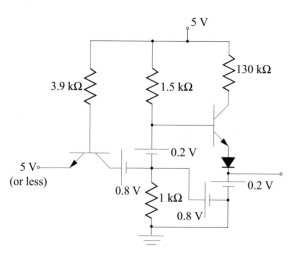

In this state, there is insufficient voltage between the base of Q_4 and the output terminal to forward bias both the base-emitter junction of Q_4 and the diode. Therefore, Q_4 in the master gate is OFF. In order to determine the ratio of collector to base currents in the output transistor, Q_3, the collector current of Q_2 must be calculated:

$$I_{C2} = \frac{5 - 0.8 - 0.2}{1.5\,k} = 2.667\,\text{mA}.$$

The base current of Q_3 is therefore:

$$I_{B3} = I_{B2} + I_{C2} - \frac{0.8}{1\,k} = 2.559\,\text{mA}.$$

The no-load collector current of Q_3, $I_{C3(nl)}$, in the master gate is zero due to Q_4 being in the cut-off region: thus the master gate fan-out is determined from the saturation conditions on Q_3:

$$I_{C3} < \beta_F I_{B3},$$

or

$$I_{C3(nl)} + N(I_{in}) < \beta_F I_{B3},$$

or

$$0\,\text{mA} + N(1.026\,\text{mA}) < 50(2.559\,\text{mA}) \quad \Rightarrow \quad N < 124.7.$$

The fan-out of this gate is 124 gates of similar construction. This is an increase of only one gate over the fan-out for the resistive pull-up gate discussed in Example 2.8 (Book 1)).

Additional changes in the circuit topology can bring further improvements. The circuit of Figure 16.7 shows two such topological changes. The active pull-up circuit in this TTL NAND gate realization consists of two BJTs connected as a Darlington pair[6] rather than the BJT-diode connection previously discussed. Also shown is an active pull-down circuit connected to the base of the output transistor, Q_3.

The Darlington active pair pull-up, formed by Q_4 and Q_5, presents a very low Thévenin resistance to the collector of Q_3 and further reduces the saturation overdrive factor, N_1:

$$R_{th} \approx \frac{1}{\beta_{F4} + 1} \left(R_{b4} // \frac{R_{c2}}{\beta_{F5} + 1} \right) \approx \frac{R_{c2}}{(\beta_{F4} + 1)\,(\beta_{F5} + 1)}.$$

This Darlington pull-up topology was first commercially seen in the 74HXX/54HXX high-speed series of TTL gates. The active pull-down circuit also shown in Figure 16.7 is formed by two resistors, R_{b3} and R_{b6}, and a transistor, Q_6. The effect of this configuration is to present a low Thévenin resistance to the base of Q_3:

$$R_{b3th} \approx \frac{R_{b6}}{\beta_{F6} + 1}.$$

Lowering the apparent base resistance increases the pull-down base current thereby increasing the reverse overdrive factor, N_2. Increased reverse overdrive factor lowers both the storage and fall times for Q_3.

Saturation Control

Any attempt to quickly remove charge stored in the junctions of a saturated transistor by reducing various Thévenin resistances has physical limits. The next logical step in improving gate performance is centered in limiting the charge build-up by controlling the region of operation of the

[6]Darlington pairs are discussed in Section 6.2 (Book 2).

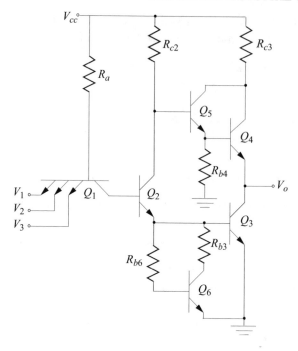

Figure 16.7: TTL gate with active pull-down and improved pull-up.

transistor. Specifically, if the transistors never enter the saturation region, the component of the storage time related to the transistor storage time constant, τ_s, will become insignificant. Since this component can easily be the greatest contributor to slow transition times, gate speed will be dramatically improved. An exclusion from the saturation region can be accomplished if the base-collector junction is shunted by a low-V_γ diode that is less subject to charge-storage effects than the transistor base-collector junction. The diode shunting the base-collector junction will not allow that junction to become sufficiently forward biased thereby keeping the BJT in the forward-active region. While a Germanium diode ($V_\gamma \approx 0.3$ V) seems ideal for the purpose of BJT saturation control,[7] fabrication of Silicon BJTs and Germanium diodes on the same IC chip is not practical. Schottky-barrier diodes are the ideal alternative.

Schottky-barrier diodes are formed with a metal-semiconductor junction rather than the usual *p-n* semiconductor junction. A representation of the metal-semiconductor junction and the circuit symbol for a Schottky-barrier diode is shown in Figure 16.8. At the junction between a metal and a semiconductor, the metal acts as a *p*-type impurity. If the semiconductor is *n*-type, the junction acts as a diode. The V-I characteristic of a Schottky-barrier diode is indistinguishable

---

[7]The shunting of the base-collector junction of a Silicon BJT with a Germanium diode is referred to as *Baker clamping* a Silicon transistor.

from that of a p-n junction except that the cut-in voltage, V_γ, is somewhat smaller. Depending on the doping of the semiconductor and the metal used, V_γ ranges between 0.2 V and 0.5 V with Silicon as the semiconductor. A typical IC realization of a Schottky-barrier diode using Aluminum results in $V_\gamma \approx 0.4$ V. One particular advantage of a Schottky-barrier diode is the extremely small charge storage time (equivalently, a small junction capacitance) associated with the junction. This storage time is typically at least an order of magnitude smaller that an equivalently sized p-n junction.

Figure 16.8: A Schottky-barrier diode.

Transistors that incorporate a Schottky-barrier diode shunting the base-collector junction are referred to as Schottky transistors. The circuit symbol for a Schottky npn transistor and a conceptual IC realization of this transistor are shown in Figure 16.9. The metallic base electrode bridges the p-type material that is the base and the n-type material that forms the collector allowing it to serve dual purposes: to be the base contact and also to form the base-collector shunting Schottky diode.[8]

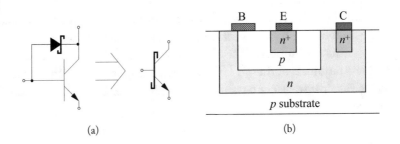

(a)   (b)

Figure 16.9: A Schottky transistor: (a) Circuit symbol; (b) IC realization.

If all saturating transistors in a TTL or DTL gate are replaced by Schottky transistors, the speed of the gate will be significantly improved. The 74SXX/54SXX series of TTL gates have the same topology as the gate shown in Figure 16.7 with all BJTs *except* Q_4 replaced by Schottky transistors (Q_4 is not a saturating transistor). Other than reduced BJT storage time, the change to Schottky transistors has little effect on the performance of the NAND gate. The greatest change is in the nominal logic levels. The HIGH output level remains unchanged at $V_{oH} \approx 3.5$ V, but the LOW level is increased slightly. Since the output transistor, Q_3, no longer saturates with a logic

---

[8]The collector region of this BJT is shown with both an n and n^+ region. This process prevents the formation of a Schottky barrier diode at the metal-semiconductor junction.

LOW output the new low output is given by:

$$V_{oL} = V_{BE3(\text{active})} - V_\gamma = 0.7 - 0.4 \approx 0.3\,\text{V}.$$

Similarly, the extremes of the input logic levels are given by:

$$V_{iH(\text{min})} = V_{BE3(\text{active})} + V_{BE2(\text{active})} + (V_{BE1(\text{active})} - V_\gamma) \approx 1.7\,\text{V},$$

and

$$V_{iL(\text{max})} = V_{BE3(\text{cut-in})} + V_{BE2(\text{active})} - (V_{BE1(\text{active})} - V_\gamma) \approx 0.9\,\text{V}.$$

The change in input and output logic levels also alters the noise margins. The noise margins for a Schottky TTL gate (74SXX) are given by:

$$\text{NM(HIGH)} = V_{oH} - V_{iH(\text{min})} = 3.5 - 1.7 \approx 1.8\,\text{V}$$
$$\text{NM(LOW)} = V_{iL(\text{max})} - V_{oL} = 0.9 - 0.3 \approx 0.5\,\text{V}.$$

Further improvements in IC fabrication techniques have allowed designers to return to a DTL input to reduce power consumption. The 74LSXX/54LSXX series of low-power Schottky gate (Figure 16.10) is an example of the return to DTL gate topology with active pull-up and pull-down improvements.

The return to DTL technology was facilitated by the Schottky transistor, Q_2. Since this BJT no longer enters the saturation region, it is not necessary to have active elements to speed the charge dissipation during logic transitions. Q_1 was one such active element: it can return to a DTL topology without significant loss of gate speed. Another common gate design feature is shown at the inputs terminals of the TTL gate shown in Figure 16.10. The inputs of this series of gates are protected against transient current surges with a reverse-biased Schottky diode shunting each input to ground.

Open-Collector Outputs

TTL gates are also available with no internal provision for either active or passive pull-up of the output transistor. Such a gate is identified as having an *open-collector output*. The two primary advantages of an open-collector output are:

- wired-AND operations can be simply created

- simple external loads are easily driven

Open-collector outputs as also useful in driving a simple load such as a LED or a relay. An example of wired-AND is shown in Figure 16.11. In this circuit, the outputs of two open-collector NAND gates are connected through an external resistor to positive power. The system output can only go HIGH if both NAND gate output BJTs are OFF. Since the output BJT OFF state is associated with a HIGH output, the logical operation is an AND operation on the outputs of the two gates.

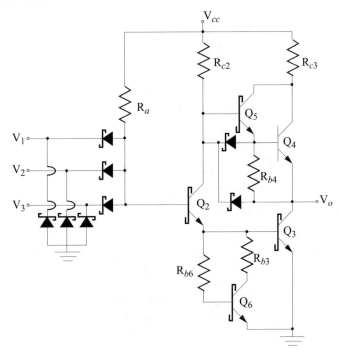

Figure 16.10: Low-power Schottky DTL (LS) gate with active pull-up and pull-down.

In order to ensure proper operation of a wired-AND, the external resistor, R_L, has a minimum value. It must be chosen so that:

$$R_L > \frac{V_{CC} - V_{OL(\text{max})}}{I_{OL(\text{max})} - N\,I_{IL}}.$$

The denominator terms are defined as: $I_{OL(\text{max})}$ is the maximum output current that the output BJT of a gate can sink ($\beta_F I_B$), $N I_{IL}$ is the current drawn by N gates with a LOW input. Gates with internal passive pull-up resistors typically violate the constraints on R_L and therefore must not be connected in this manner. If gates with active pull-up are connected in this manner, excessive power will be dissipated in the output stage of an individual gate leading to gate failure.

Open collector gates, when properly terminated, have passive pull-up: as such they exhibit relatively long propagation delays compared to gates with active pull-up.

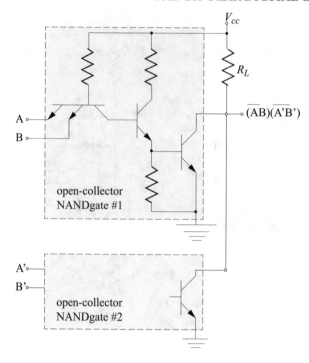

Figure 16.11: Two open-collector NAND gates with outputs forming a wired-AND.

16.2.2 ECL LOGIC GATES

The basic functional topology of all ECL gates is shown in Figure 16.12. All ECL gates use this basic topology only with variations in the individual components. The current switch is always in the form of several BJTs sharing a common connection at the emitter terminal. This configuration, essentially a very high-gain differential amplifier, ensures that the emitter-coupled BJTs never enter the saturation region: switching is between the forward-active region and cut-off. As such, ECL gates have a short storage time and are inherently the quickest form of bipolar logic. The bias network provides stable voltage references and the output driver buffers the output of the current switch to increase fan-out and match impedance for optimal transmission.

The significant advantages of the ECL gate family over other bipolar logic families include:

- Complementary outputs. Most logic elements offer both the logic function and its complement (i.e., OR/NOR). Additional logic inverters are eliminated from designs reducing timing delays and power consumption.

- Constant supply current. The current drain remains essentially constant regardless of gate logic state. The power supply design requirements are therefore simplified.

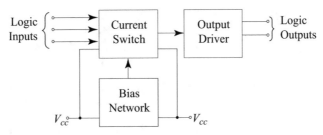

Figure 16.12: Basic ECL gate topology.

- Low switching noise. The differential amplifier and common-collector amplifier switching sections of ECL logic have very low current transients. In addition, the low voltage variation in logic states ensures that charging of any stray capacitance results in small current variation.

- Low crosstalk. Crosstalk in digital systems is proportional to the derivative of voltage signals. Low logic-level voltage differences, as well as greatly reduced BJT storage times, allow for effective control of signal rise and fall times without significantly slowing gate speed. Voltage rise and fall rates can be reduced in ECL by a factor of five (or more) over typical TTL rates.

- High fan-out. The pairing of high input impedance with low output impedance allows for large fan-out.

The disadvantages are primarily higher power consumption, lower noise margins, and additional design constraints in the IC environment. ECL is operated with a negative power supply to reduce the effects of noise and power supply variation on gate output.

The basic operation of an early ECL OR gate[9] was described in Section 3.5 (Book 1). In that basic circuit the bias network is an externally supplied reference voltage. Also, the output common-collector buffer has a pull-down resistor.[10] Later ECL gate families incorporate several improvements. One series of gates that incorporates the significant changes in ECL topology is the ECL 10K series. The topology of an ECL 10K OR/NOR gate is shown in Figure 16.13. The circuit topology of the 10K series differs from the basic gate in several ways:

- Internal bias network. A bias network is included in the IC realization of the gate to provide a reference voltage to the differential amplifier.

- Ground separation. Two separate ground terminals are provided so that power supply transients due to switching in the differential amplifier are not passed through the ground to the output buffers.

[9]The gate described in Section 3.5 (Book 1) has the topology of a MECL I circuit. MECL II has identical topology except the bias network is included in the IC realization.
[10]A pull-down resistor at the emitter of the output BJT ensures that the output will appropriately be pulled LOW.

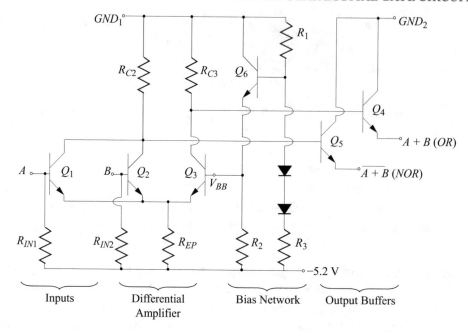

Figure 16.13: Basic ECL 10K OR/NOR gate.

- Relocation of pull-down. The relatively low-value, output-emitter, pull-down resistors have been eliminated and replaced by high-value, resistors on the input BJTs. Benefits include reduced power consumption, more reliable gate operation with floating inputs, and increased gate life.

Logic levels remain unchanged at:

$$V_{oH} \approx -0.72\,\text{V} \quad \text{and} \quad V_{oL} \approx -1.60\,\text{V}.$$

One problem that is common in ECL gates is the variation of the logic levels due to variation in the power supply voltage or due to temperature changes. Various advanced ECL circuits have addressed these issues at the expense of circuit complexity. One such advanced circuit topology is that of the ECL 100K series shown in Figure 16.14. Notice that the increase in power consumption due to added circuitry has been compensated for by a reduction in the power supply voltage magnitude from 5.2 V to 4.5 V.

16.2.3 I²L LOGIC GATES

While TTL and ECL circuits are a good choice for small-scale (SSI) and medium-scale (MSI) integrated circuits, they have limited use in large-scale (LSI) and are not practical for very large-

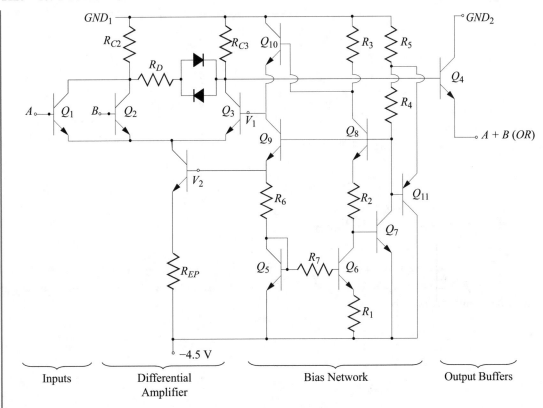

Figure 16.14: ECL 100K OR gate.

scale (VLSI) applications. The primary reasons for these limitations on the use of TTL and ECL circuits are:

- Relatively high power consumption, and

- Relatively large surface area of IC realization.

Integrated injection logic[11] (I^2L) blends high speed with high surface area density and low power consumption. In addition, the simple gate structure of I^2L provides multifunction outputs. For example, the simple two-input OR/NORgate shown in Figure 16.15 provides four logical operations as outputs: the complement of each input, $\bar{A}$ and $\bar{B}$, and the OR/NOR functions, $A + B$ and $\overline{A + B}$.

[11]I^2L is also known as *merged transistor logic* (MTL).

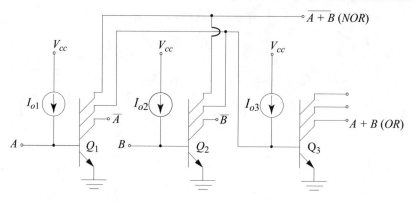

Figure 16.15: Simple I^2L OR/NOR logic gate.

The basic element of an I^2L gate is a multiple-collector *npn* BJT driven at the base with a simple *pnp* BJT current source (Figure 16.16). The unique feature of this element is its single input and multiple outputs.

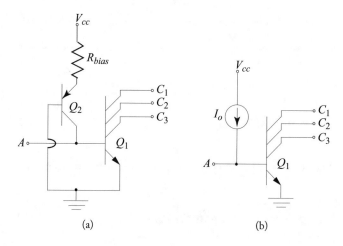

Figure 16.16: Basic I^2L digital gate element: (a) Circuit diagram; (b) Equivalent circuit.

The operation of the basic I^2L element is simple if it is remembered that at least one of the multiple collector outputs must be connected to the input of another element in a master/slave pairing. If the master gate input, A, is HIGH, Q_1 enters the saturation region and the master gate output at each of the multiple collectors is $V_{CE(sat)}$. If the master gate input is LOW, Q_1 enters the cut-off region. The slave gate sees a high impedance input from the master gate. The current source forces the slave gate Q_1 into saturation which, in turn, forces the slave gate input *and* the master gate output to $V_{BE(sat)}$. All other collectors of the master gate will also be raised to that voltage level. The output logical voltage levels are therefore:

$$V_{oH} = V_{BE(sat)} \approx 0.8 \, \text{V} \quad \text{and} \quad V_{oL} = V_{CE(sat)} \approx 0.2 \, \text{V}.$$

The input logical levels are the levels at which a BJT can be considered to be ON or OFF:

$$V_{iH(min)} = V_{\gamma} \approx 0.7 \, \text{V} \quad \text{and} \quad V_{iL(max)} = V_{BE(on)} \approx 0.6 \, \text{V}.$$

These voltage levels are all within ~ 0.6 V with the transition region only ~ 0.1 V wide. Obviously I^2L gates are not appropriate for use in a noisy environment. The noise margins are:

$$\text{NM(HIGH)} = V_{oH} - V_{iH(min)} = 0.8 - 0.7 \approx 0.1 \, \text{V}$$
$$\text{NM(LOW)} = V_{iL(max)} - V_{oL} = 0.6 - 0.2 \approx 0.4 \, \text{V}.$$

Fan-out considerations are particularly simple to visualize in I^2L gates. A HIGH output implies no interaction between master and slave gate: there is no limit on the number of slave gates imposed by a logic HIGH output. For a LOW output, the master gate output BJT must be in saturation and the slave gate output BJT in cut-off. The slave gate therefore supplies I_o to the collector of the master gate output BJT. The condition for saturation is:

$$\beta_F I_B \geq I_C \quad \Rightarrow \quad \beta_F I_o \geq N I_o \quad \Rightarrow \quad \text{fan-out} = \beta_F.$$

It must be noted that the physical structure of the multiple-collector BJT limits the forward current gain, β_F. In most cases β_F for a multiple-collector BJT is much smaller than that of modern, simple BJTs: consequently, *the fan-out of an I^2L gate is typically less than ten gates of similar construction.* If the output to the final gate of a logical operation is to be taken off the chip, it is necessary to passively pull-up the collector of the output BJT. This can be accomplished with a resistor connected between any of the output BJT collector terminals and a positive voltage supply.

The interconnection of the two BJTs that make up an I^2L gate simplifies IC realization of the gate structure. The p-type base of the current source is directly connected to the p-type emitter of the inverter: similarly, the n-type collector of the current source connects to the n-type base of the inverter. These interconnections lead to shared regions in the IC realization as shown in Figure 16.16. A single resistor is usually used for all I^2L gates on a chip. A multiple collector structure can be constructed in a similar fashion to the multiple emitter structure of the input transistor of TTL gates.

16.3 DYNAMIC PROPERTIES OF METAL OXIDE SEMICONDUCTOR TRANSISTORS

Metal Oxide Semiconductor (MOS) Digital Logic gates depend on the transition of the output of a MOSFET from one logic level to another. As in bipolar transitions, the speed at which

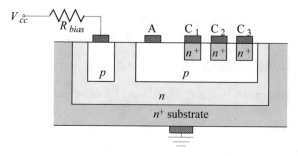

Figure 16.17: IC realization of an I^2L gate with three collectors.

any gate operates is limited by the physical parameters of its constituent transistors and of the components in the circuit surrounding the transistor. While it is impossible to analyze the effect of all possible circuit topologies on switching speed, an analysis of the switching speed the simple CMOS inverter provides much insight into all MOS switching.[12] As shown in Figure 16.18, a general CMOS inverter consists of two *complementary* MOS transistors: the input to the inverter is at the connection of the two FET gate terminals and the output is at the connection of the drain terminals. Here it is assumed that the master inverter shown drives a similar-topology slave inverter. This slave inverter is represented by its input impedance: a capacitor in parallel with a very large resistance. As the input resistance of a FET is essentially infinite, it is assumed to have no significant effect on any further calculations.

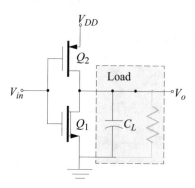

Figure 16.18: Typical CMOS inverter, capacitively loaded.

The response of the CMOS inverter to a logic LOW to HIGH input transition is shown in Figure 16.19. For a CMOS inverter these logic levels are:

$$V_{iL} \approx 0\,\text{V} \quad \text{and} \quad V_{iH} \approx V_{DD}.$$

[12]This digital CMOS inverter is initially discussed in Section 3.4 (Book 1).

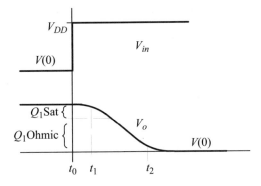

Figure 16.19: Transient response of a CMOS inverter.

As in the case of a bipolar inverter, the digital transition can not be instantaneous and must experience some delay. Due to the symmetry of the CMOS circuit, the *mathematical expressions* for the rise and fall time will be functionally identical. The rise and fall times themselves will vary with the transistor characteristic. Specifically, the rise and fall times are dependent on the *n*-channel and *p*-channel FET transconductance parameters, K_n and K_p, respectively, as well as the respective threshold voltages.

The response to a digital step consists of regions of constancy and a region of transition. In the region of transition there are two significant time periods.

- Delay Time $= t_d = t_1 - t_0$

- Rise Time $= t_r = t_2 - t_1$

The delay time is the time between the input pulse transition and when the response transitions 10% of the distance between HIGH and LOW states. The rise time is the time for a 10% to 90% transition from HIGH to LOW. The term *rise time* is used to match the definitions in bipolar circuitry. In NMOS gate circuits, it refers to the FET drain current change: as the voltage transitions from a HIGH to LOW, the drain current rises from a minimum to a maximum value. It is a misnomer for CMOS circuits: the drain current is zero for both logic states.

During the delay time ($t_o \leq t < t_1$), the *n*-channel MOSFET is in the saturation region of operation and is described by the voltage-current relationship:[13]

$$I_D = K(V_{GS} - V_T)^2. \tag{16.24}$$

Since the input voltage to the inverter is a constant value ($V_{in} = V_{DD}$) during this time period, it can be seen through Equation (16.24), that the drain current is constant during the delay time.

---

[13]The voltage-current relationships for FETs of all types are presented in Table 4.2 (Book 1).

The load capacitance can therefor be assumed to discharge linearly:

$$I_D \Delta t = -C_L \Delta V_o \quad \Rightarrow \quad t_d = \frac{-C_L \Delta V_o}{I_D}. \tag{16.25}$$

The change in the output voltage, ΔV_o, is 10% of the supply voltage: $\Delta V_o = -0.1 V_{DD}$. The total delay time can therefore be easily computed:

$$t_d = \frac{0.1 \, C_L V_{DD}}{K \, (V_{DD} - V_T)^2}. \tag{16.26}$$

The rise time has two components:

$$t_r = t_{r1} + t_{r2}, \tag{16.27}$$

where t_{r1} is the portion of the rise time where Q_1 is in the saturation region and t_{r2} is the portion where Q_1 is in the ohmic region. The transition between the saturation and ohmic regions of a enhancement region MOSFET occurs when:

$$V_{DS} = V_{GS} - V_T \quad \Rightarrow \quad V_o = V_{DD} - V_T. \tag{16.28}$$

As during the delay time, the drain current is constant for a saturation region FET and the load capacitor discharges linearly during t_{r1}:

$$t_{r1} = \frac{-C_L \Delta V_o}{I_D}, \tag{16.29}$$

or

$$t_{r1} = \frac{-C_L \left((V_{DD} - V_T) - 0.9 V_{DD} \right)}{K \, (V_{DD} - V_T)^2} = \frac{C_L \, (V_T - 0.1 \, V_{DD})}{K (V_{DD} - V_T)^2}. \tag{16.30}$$

During the second portion of the rise time, Q_1 is in the ohmic region and can be described by the expression:

$$I_D = K \left[2 \, (V_{GS} - V_T) \, V_{DS} - V_{DS}^2 \right]. \tag{16.31}$$

The load capacitor discharge is described by the differential form of its voltage-current relationship:

$$I_D = -C_L \frac{d V_o}{dt}. \tag{16.32}$$

Simultaneously solving Equations (16.31) and (16.32) leads to an integral expression for t_{r2}:

$$t_{r2} = \frac{C_L}{2K \, (V_{DD} - V_T)} \int_{V_{DD} - V_T}^{0.1 V_{DD}} \frac{d V_o}{\left[\dfrac{V_o^2}{2 \, (V_{DD} - V_T)} - V_o \right]}. \tag{16.33}$$

Evaluation of the integral gives the expression for the second portion of the rise time:

$$t_{r2} = \frac{C_L}{2K(V_{DD} - V_T)} \ln \left\{ 20\frac{V_{DD} - V_T}{V_{DD}} - 1 \right\}. \tag{16.34}$$

The total rise time for the CMOS inverter is the sum of the rise time components:

$$t_r = \frac{C_L}{K(V_{DD} - V_T)} \left[\frac{(V_T - 0.1V_{DD})}{(V_{DD} - V_T)} + \frac{1}{2} \ln \left\{ 20\frac{V_{DD} - V_T}{V_{DD}} - 1 \right\} \right]. \tag{16.35}$$

A comparison of the expressions for delay time (Equation (16.26)) and the rise time (Equation (16.13)) shows that the rise time dominates the delays inherent in digital CMOS switching. In addition, both the rise time and the delay time are directly proportional to the capacitance of the load. This dependence on load capacitance is the determining factor in MOS gate fan-out. When many slave gates are connected to the output of a single master gate, the input capacitance of the slave gates add. The rise and delay times increase directly in proportion to the number of slave gates attached to the output of the master gate. Transition speed requirements put an upper limit on this number.

As with the bipolar inverter, propagation delay is a useful descriptor of the gate speed. The propagation delay of a CMOS inverter can be described in terms of the transition times derived. Its component terms are:

$$t_{PHL} \approx t_d + \frac{t_r}{2} \qquad (n\text{-channel FET parameters}). \tag{16.36}$$

The LOW to HIGH propagation time has the same mathematical form but the transconductance parameter of the p-channel FET must be used.

$$t_{PLH} \approx t_d + \frac{t_r}{2} \qquad (p\text{-channel FET parameters}). \tag{16.37}$$

The average propagation time is given by:

$$t_P = 1/2(t_{PHL} + t_{PLH}). \tag{16.38}$$

Example 16.2

A CMOS inverter is fabricated using a 5 V supply and MOSFETs with the following properties:

$$K_N = 0.1 \, \text{mA/V}^2 \ (n\text{-channel}) \qquad |V_T| = 1$$
$$K_P = 0.25 \, \text{mA/V}^2 \ (p\text{-channel}).$$

Determine the average propagation delay time if it is driving a capacitive load of 5 pF.

Solution:

Equation (16.26) yields the delay times:

$$t_d = \frac{0.1 C_L V_{DD}}{K(V_{DD} - V_T)^2} = \frac{0.156 \times 10^{-12}}{K}.$$

The HIGH to LOW transition uses K_N while the LOW to HIGH transition uses K_P

$$t_{dHL} = 1.56 \text{ ns} \qquad t_{dLH} = 0.625 \text{ ns}.$$

Similarly, the rise times can be calculated from Equation (16.35):

$$t_r = \frac{1.849 \times 10^{-12}}{K}$$

$$t_{rHL} = 18.49 \text{ ns} \qquad t_{rLH} = 7.40 \text{ ns}.$$

The individual propagation delays are given by

$$t_{PHL} \approx t_{dHL} + 1/2 t_{rHL} = 10.81 \text{ ns}$$

$$t_{PLH} \approx t_{dLH} + 1/2 t_{rLH} = 4.32 \text{ ns}.$$

The average propagation delay is the average of the individual propagation delays:

$$t_P = 1/2(t_{PHL} + t_{PLH}) \approx 7.56 \text{ ns}.$$

An NMOS inverter is similar to a CMOS inverter but fabricated with an n-channel active load rather than the p-channel switch. Transition time calculations are similar to those described for the CMOS inverter, with the exception that the switching transistor, Q_1, is always in the ohmic region. The expansion of NMOS inverter calculations to cover NMOS gates is complicated further due to the dependence of the LOW output voltage, V_{oL}, on the exact state of the multiple switching transistors inherent to MOS gates. CMOS and NMOS gates operate at *essentially* the same speed if comparable FETs are used.

FET Latch-up

The IC realization of MOS gate structures produces, in addition to the MOS structures, parasitic bipolar structures. While the *npn* and *pnp* structures that produce parasitic BJTs are usually benign, *pnpn* structures can produce a parasitic silicon-controlled rectifier (SCR).[14] An SCR is a form of latching switch that is activated by the proper injection of a small current or by a high-derivative voltage pulse. Once the SCR structure is activated, the high currents that result, combined with the latching property of the SCR, lead to catastrophic failure of the MOS device. Modern IC design of MOS structures includes transient protection structures so that the possibility of parasitic SCR latch-up is minimized in all but the noisiest of environments. Still, MOS ICs are particularly sensitive to damage by static discharge.

[14] SCRs are discussed extensively in Section 14.1.

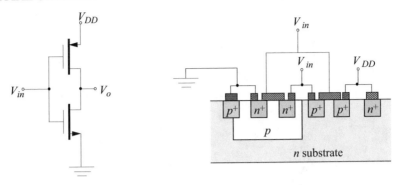

Figure 16.20: CMOS inverter and conceptual IC realization.

16.4   MOS DIGITAL GATE CIRCUITS

The two basic logic operations in MOS gates are NAND and NOR. In order to create these logic operations, MOSFETs are used essentially as switches. The two logic operations can be conceptualized (Figure 16.21) as a resistive load in series with either a series connection of several switches to ground (NAND) or a parallel connection of several switches to ground (NOR). In the NAND configuration, the output will be LOW only if both of the series switches are closed: in the NOR configuration, the output is LOW if either of the parallel switches is closed. In MOS realizations of such logic gates, the FET can be used for all components of the gate: the switches as well as the active resistive loads. The extremely low current draw at the controlling terminal of a FET switch, the gate terminal, facilitates cascading these simple logic gates without many of the problems caused by loading.

MOS logic gate circuits can be constructed using MOS transistors of the same type or transistors of mixed types. NMOS circuits use n-channel FETs exclusively, albeit often with a mix of depletion and enhancement mode FETs. PMOS is the complementary form of NMOS using p-channel FETs only. CMOS (complementary MOS) uses a mixture of complementary, enhancement mode, n-channel and p-channel FETs. Of these three types, NMOS and CMOS are the most common: discussion here is restricted to the fundamental gates common to NMOS and CMOS logic.

NMOS Gates

NMOS gates are used extensively in LSI and VLSI microprocessors, memories, and other circuitry, but are not commonly available as individually packaged circuits. All elements of basic NMOS logic gates are fabricated from n-channel FETs: the switches are enhancement-mode NMOS FETs and the active resistive load can be either an enhancement mode or depletion-mode NMOS FET. The logic gate action is essentially the same for each type of active load: for simplicity, discussion here will be limited to enhancement type active loads. The circuit diagrams

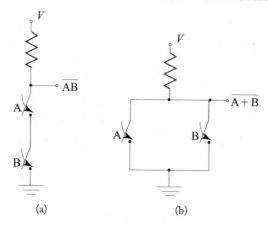

Figure 16.21: Two-input logic gates using simple controlled switches: (a) NAND; (b) NOR.

for basic, two-input NMOS NAND and NOR gates, using enhancement mode active loads, are shown in Figure 16.22. It should be noted that the geometry of all the switching FETs is identical, but the active load FET usually has different characterizing parameters.

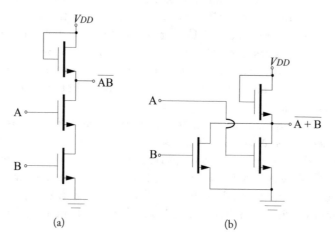

Figure 16.22: Two input NMOS logic gates (enhancement mode active load): (a) NAND; (b) NOR.

While the fundamentals of NMOS gate operation are fairly easily understood, the specifics can be more complicated. Analysis of a NOR gate is mathematically the least complicated: discussion will begin with the NOR gate and the progress to the NAND gate. The short-circuit connection between the drain and gate terminals of the active load ensures that the load FET is always in the saturation region of operation: $V_{GS} = V_{DS}$. The equation that relates the load FET drain current to its gate-source voltage is found in Table 4.2 (Book 1):

$$I_{DR} = K_R(V_{GSR} - V_T)^2. \tag{16.39}$$

An additional subscript is added to the FET parameters to specify the FET under consideration: "R" indicates the active resistive load and "S" indicates the switching FETs.[15]

In a two-input NOR gate, two LOW inputs produce simple results. For the switching FETs, $V_{GS} < V_T$ implies that the drain current, I_{DS}, is zero valued. Consequently, the drain current of the load FET must be zero valued. Substitution into Equation (16.39) yields:

$$0 = K_R(V_{GSR} - V_T)^2 \quad \Rightarrow \quad V_{GSR} = V_T. \tag{16.40}$$

The output of the gate goes to a HIGH state, V_{oH}:

$$V_{oH} = V_{DD} - V_{GSR} = V_{DD} - V_T. \tag{16.41}$$

One or more HIGH inputs to the NOR gate implies that some current flows through the FETs. Any switching FET with a HIGH input is forced into the ohmic region. Solving for currents and voltages requires the use of additional FET characteristic equations. The current characteristic equation for enhancement mode FETs operating in the ohmic region is also found in Table 4.2 (Book 1):

$$I_{DS} = K_S \left[2 \left(V_{GSS} - V_T \right) V_{DSS} - V_{DSS}^2 \right]. \tag{16.42}$$

Since there exists the possibility for N switching FETs to be simultaneously ON, the relationship between switch and load drain currents is given by:

$$I_{DR} = N I_{DS}, \tag{16.43}$$

or

$$K_R(V_{GSR} - V_T)^2 = N K_S \left[2 \left(V_{GSS} - V_T \right) V_{DSS} - V_{DSS}^2 \right]. \tag{16.44}$$

Kirchhoff's laws relate V_{GSR} to V_{DSS}:

$$V_{GSR} = V_{DD} - V_{DSS}. \tag{16.45}$$

Equation (16.44) becomes a quadratic equation in the LOW output voltage level, $V_{oL} = V_{DSS}$: as a function of the number of HIGH inputs, N, the voltage supply, V_{DD}, the FET threshold voltage, V_T, the FET transconductance factors, K_R and K_S, and the input voltage, V_{GSS}:

$$K_R(V_{DD} - V_{DSS} - V_T)^2 = N K_S \left[2 \left(V_{GSS} - V_T \right) V_{DSS} - V_{DSS}^2 \right]. \tag{16.46}$$

[15]It is assumed in this discussion that the threshold voltage, V_T, is the same for all FETs in the circuit. In IC fabrication this is a realistic design assumption.

The quadratic functional form of Equation (16.46) obscures intuition. Of most significance is a decrease in the output voltage, V_{DSS}, as NK_S/K_R increases. For most designs, it is important to set a maximum value on this LOW output. The maximum value will occur for only one switch ON ($N = 1$). A particular design goal will the restrict the minimum ratio of transconductance factors, K_S/K_R. The usual implication of this restriction is switching FETs that are significantly wider than the load FET.

Example 16.3   Design Example

Given a 5 V power supply. Design a three-input NOR gate to have the following output logic levels when driven by a gate of the same design:

$$V_{oH} = 3.8\,\text{V} \qquad\qquad V_{oL(\text{max})} = 0.9\,\text{V}.$$

Solution:

An NMOS circuit topology similar to Figure 16.22b but with three inputs and three switching FETs will satisfy the design requirements. All that remains to be determined is an acceptable set of FET parameters, V_T, K_R and K_S.

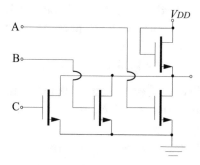

The HIGH output level is given by Equation (16.41):

$$V_{oH} = V_{DD} - V_T,$$

or

$$3.8 = 5 - V_T \quad\Rightarrow\quad V_T = 1.2\,\text{V}.$$

The maximum LOW output level is determined with only one switching FET on. For this condition the input voltage $V_{GSS} = V_{oH}$ and the output voltage $V_{DSS} = V_{oL(\text{max})}$:

$$K_R(5 - 0.9 - 1.2)^2 = 1K_S[2(3.8 - 1.2)(0.9) - (0.9)^2],$$

or

$$K_R = 0.460K_S \quad\text{or more exactly}\quad K_R \leq 0.460K_S.$$

In order to keep drain currents reasonable, the value of the transconductance factors must be kept reasonably small. Since there are no specific design constraints on FET transconductance factors, a drain current of approximately 1.93 mA will result in:

$$K_R = 230\,\mu A/V^2 \quad \text{and} \quad K_S = 500\,\mu A/V^2.$$

Other similar values will produce acceptable results. The LOW output logic level can be found to be: 0.9 V for a single HIGH input, 0.527 V for 2 HIGH inputs, and 0.373 V for 3 HIGH inputs.

The analysis of an NMOS NAND gate is similar to that of the NOR gate but complicated by the series connection of the switching FETs. The HIGH output voltage is determined in the same manner as for a NOR gate and produces the same result:

$$V_{oH} = V_{DD} - V_T. \tag{16.47}$$

The series connection of switching FETs adds complexity to the LOW output voltage calculation. While the gate-source voltage, V_{GS1S}, of the grounded-source FET is simply the input voltage, subsequent switching FETs have V_{GS} reduced by the drain-source voltages of any intervening FETs. In the two-input NAND gate circuit of Figure 16.22a, the upper switching FET has a gate-source voltage (assuming the gate is driven by a similar gate):

$$V_{GS2S} = V_{in} - V_{DS1S} = V_{DD} - V_T - V_{DS1S}, \tag{16.48}$$

while the grounded-source switching FET has gate-source voltage:

$$V_{GS1S} = V_{in} = V_{DD} - V_T. \tag{16.49}$$

Of course the NAND gate output voltage for a logic LOW is given by the sum of the switching FET drain-source voltages:

$$V_{oL} = \sum_{i=1}^{N} V_{DSiS} = V_{DS1S} + V_{DS2S}, \tag{16.50}$$

where N is the number of series switching FETs. The series connection of N identical switching FETs creates N equations relating the drain currents:

$$K_R (V_{DD} - V_{oL} - V_T)^2 = K_S \left[2 (V_{GSiS} - V_T) V_{DiSS} - V_{DiSS}^2 \right] \tag{16.51}$$
$$i = 1, 2, \ldots, N.$$

The simultaneous solution of Equations (16.48) and (16.49) determines the LOW output voltage level.

Example 16.4   Design Example

Given a 5 V power supply. Design a three-input NAND gate to have the following output logic levels when driven by a gate of the same design:

$$V_{oH} = 3.8\,\text{V} \qquad\qquad V_{oL(\text{max})} = 0.5\,\text{V}.$$

Solution:

An NMOS circuit topology similar to Figure 16.22a but with three inputs and three switching FETs will satisfy the design requirements. All that remains to be determined is an acceptable set of FET parameters, V_T, K_R and K_S.

The HIGH output level is given by Equation (16.41)

$$V_{oH} = V_{DD} - V_T,$$

or

$$3.8 = 5 - V_T \quad\Rightarrow\quad V_T = 1.2\,\text{V}.$$

The maximum LOW output level is determined by simultaneously solving Equations (16.50) and (16.51) (four total equations):

$$0.5 = V_{DS1S} + V_{DS2S} + V_{DS3S}$$
$$K_R(V_{oL} - 1.2)^2 = 1K_S[2(3.8 - 1.2)(V_{DS1S}) - (V_{DS1S})^2]$$
$$K_R(V_{oL} - 1.2)^2 = 1K_S[2(3.8 - 1.2 - V_{DS1S})(V_{DS2S}) - (V_{DS2S})^2],$$

and

$$K_R(V_{oL} - 1.2)^2 = 1K_S[2(3.8 - 1.2 - V_{DS1S} - V_{DS2S})(V_{DS3S}) - (V_{DS3S})^2].$$

There are no easy, closed-form techniques for solving 4 non-linear, simultaneous equations. A MathCAD solution to find the ratio of the transconductance factors follows:

MathCAD Solution for the ratio of FET transconductance factors for three input NAND gate

$V_T := 1.2 \qquad V_{oL} := 0.5$

Guess values for the ratio and the three drain-source voltages

$\text{ratio} := 1 \qquad V1 := .2 \qquad V2 := .2 \qquad V3 := .3$

Given

$V1 + V2 + V3 = V_{oL} \qquad\qquad (V_{oL} - V_T)^2 = \text{ratio}\,[2 \cdot (3.8 - 1.2 - V1) \cdot V2 - V2^2]$

$(V_{oL} - V_T)^2 + \text{ratio}\,[2 \cdot (3.8 - 1.2)\,V1 - \quad V1^2] \quad (V_{oL} - V_T)^2 = \text{ratio}\,[2 \cdot (3.8 - 1.2 - V1 - V2) \cdot V3 - V3^2]$

$\text{Find}\,(V1, V2, V3, \text{ratio}) = \begin{bmatrix} 0.155 \\ 0.166 \\ 0.179 \\ 0.626 \end{bmatrix}$

The minimum ratio of transconductance factors for a 0.5 V LOW output voltage is therefore:

$$\frac{K_S}{K_R} \geq 0.626.$$

Choosing the same load FET K_R as Example 16.3, implies that the NAND gate will draw the same power for a LOW output as the NOR gate in that example. This design choice leads to:

$$K_R = 230\,\mu A/V^2 \quad \text{and} \quad K_S \geq 144\,\mu A/V^2.$$

Choosing the same switch FET K_S as Example 16.3 reduces the output LOW voltage to $V_{oL} \approx 0.25$ V, and greatly simplifies IC layout. For these reasons, a reasonable choice might be:

$$K_S = 500\,\mu A/V^2.$$

Other choices will fulfill the design goals adequately. For example, identical FETs results in $V_{oL} \approx 0.4$ V.

As was previously stated, the active load for an NMOS gate can be either an enhancement mode or depletion mode FET. When NMOS gates use a depletion-mode load, the gate and source terminals of the active load are shorted together: this connection ensures that the depletion-mode load is always in the saturation region of operation. In the depletion-mode case, the IC fabrication procedures are more complex but typically lead to faster switching speeds. Analysis techniques, while not discussed here, are similar to those of the enhancement-mode active load case.

Rather than switches mixed with active, resistive loads, it is possible to produce NAND and NOR logic operations using only controlled switches. In this type of realization, the resistive load is replaced by a group of *oppositely* controlled switches. As shown in Figure 16.23, each input simultaneously produces an action on one of the positively controlled switches and the opposite action on its counterpart in the negatively controlled switches. CMOS logic gates are founded on this realization of logical switching.

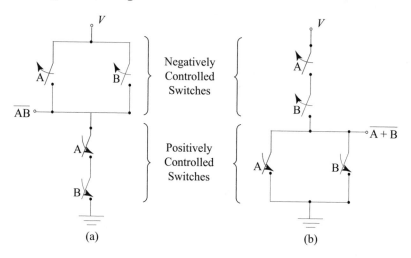

Figure 16.23: Conceptual switch-only two-input logic gates: (a) NAND; (b) NOR.

CMOS Gates

CMOS logic gates are available in SSI packages and are found in many LSI and VLSI applications such as calculators and watches. The very popular 74HCXX series of logic gates is an example of SSI CMOS logic. As the physical scale of IC circuit realization becomes smaller, CMOS is becoming the most significant form of MOS gate in VLSI applications. Part of this rise in CMOS circuitry is due to its very low power consumption.

In standard CMOS each input is connected to an individual NMOS FET *and* a PMOS FET. The complementary channel FETs act as opposite-acting switches with this connection. A logic HIGH signal closes the NMOS switch and opens the PMOS switch: a LOW signal produces the reverse actions. An example of this dual connection, a two-input NAND gate, is shown in Figure 16.24a. This NAND gate consists of two NMOS transistors in series connected to two PMOS transistors in parallel. Whenever an input turns one of the NMOS FETs ON, the corresponding PMOS FET is turned OFF (i.e., Q_{n1} ON implies Q_{p1} OFF). Thus, the output is switched to its LOW state only if *both* NMOS transistors are ON (both inputs HIGH), but connected to V_{DD} when *either* PMOS in ON (either input LOW). This is the ideal form for a NAND gate. A two-input NOR gate is the exact dual structure of the NAND gate: there are two parallel NMOS FETs transistors connected to two series PMOS FETs, as shown in Figure 16.24b. Each additional input adds two FETs: one of each type connected in series or parallel, as appropriate, to its matching-type FETs.

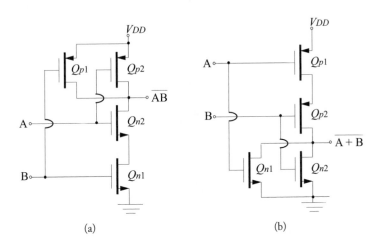

(a)   (b)

Figure 16.24: Two input CMOS logic gates: (a) NAND; (b) NOR.

CMOS gates are characterized by very stable logic voltage levels. Each FET switches between cut-off and ohmic regions operation. In the cut-off region of operation, no drain current flows through the FET. In the CMOS gate, ohmic region FETs are, in all stable states, in series with a cut-off FET (or combinations of cut-off FETs) and must also have no drain current. Consequently, CMOS gates consume power only during the switching transient. During this transient, a short-duration current pulse flow through the FETs, leading to low total power consumption. Unfortunately, it also generates significant electrical noise.

The transfer characteristic of a CMOS gate can be approached by analyzing the states of the individual FETs. An ohmic region FETs is described by:

$$I_D = K_S \left[2 \left(V_{GS} - V_T \right) V_{DS} - V_{DS}^2 \right]. \tag{16.52}$$

If the drain current must be zero-valued, the implication of this simple application of Kirchhoff's current law is:

$$V_{DS} = 0.$$

The logic voltage levels of a CMOS gate are therefore limited by the supply voltage and ground:

$$V_{oH} = V_{DD} \quad \text{and} \quad V_{oL} = 0.$$

The voltage at which the logic transitions occur is not easily characterized for CMOS gate circuits. It is a function of the p-channel and n-channel FET transconductance factors, K_P and

K_N, as well as which inputs are in transition. A typical transfer relationship for a two-input CMOS NAND gate is shown in Figure 16.25. When only input A is in transition, the transition occurs at the lowest voltage: only B in transition occurs at a slightly higher input voltage. When both inputs are simultaneously in transition, the transition occurs at its highest voltage level.

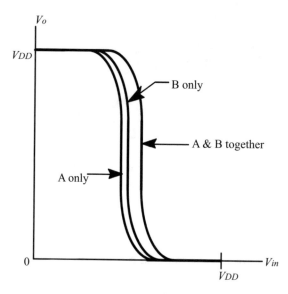

Figure 16.25: Two-input CMOS NAND gate transfer relationship.

The spread in the transition region can be investigated by observing the midpoint of the output voltage transition ($V_o = 1/2V_{DD}$) for the two extreme cases. For simplicity of discussion, assume the output is in transition from logic LOW to HIGH.

For a logic LOW, both inputs to a NAND gate are high and the FETs are in the following regions of operation:

- Q_{n1} and Q_{n2} in ohmic region

- Q_{p1} and Q_{p2} in cut-off

The lowest voltage transition occurs when the input voltage, V_{in}, is connected to terminal A. At the midpoint of the transition for this case, Q_{n2} and Q_{p2} enter the saturation region while the two other FETs remain in their previous states. The currents through the FETs are:

$$I_{Dp2} = -K_p(V_{in} - V_{DD} + V_T)^2 \tag{16.53}$$

$$I_{Dn2} = K_n(V_{in} - V_{DS1} - V_T)^2 \tag{16.54}$$

$$I_{Dn1} = K_n\{2(V_{DD} - V_T)V_{DS1} - (V_{DS1})^2\}. \tag{16.55}$$

Simultaneous solution of these three equations, knowing that the currents must have the same magnitude, leads to the value of the input voltage, V_{in}, at which the transition occurs. If both inputs transition at the same time, Q_{p1} is also in the saturation region at the midpoint of the transition. This difference in FET state leads to an increased n-channel drain current since both p-channel FETs are contributing to the total current.

As with NMOS gates, CMOS gates are optimized for size and speed as well as voltage levels. Often these design goals lead to differently characterized (and sized) n-channel and p-channel FETs.

Example 16.5

Determine the range of input voltages at which the midpoint of a logic transition occurs for a two-input CMOS NAND gate that uses FETs described by:

$$|V_T| = 1\,\text{V} \quad \text{and} \quad K_n/K_p = 2,$$

and a 5 V power supply.

Solution:

The midpoint of the logic transition is found by the simultaneous solution of Equations (16.53) through (16.55). There are actually three input voltages at which the midpoint of the transition occurs: each depends on which input is in transition. Simultaneous solution of several non-linear equations is best accomplished numerically. A MathCAD solution for two cases (input A only and both A and B) is shown at the right. The simultaneous change is described by noting that both p-channel FETs have the same terminal voltages applied.

Two-input CMOS NAND gate transition solutiom

$V_{dd} := 5 \qquad V_t := 1$

$K_n := 2 \qquad K_p := 1$

Guess Values

$V_{in} := 2 \qquad V_{ds} := .1$

Given   (Solve Block)

$(V_{in} - V_{ds1} - V_t)^2 = 2 \cdot (V_{dd} - V_t) \cdot V_{ds1} - V_{ds1}^2$

$K_n \cdot (V_{in} - V_{ds1} - V_t)^2 = N \cdot K_p \cdot (V_{in} - V_{dd} + V_t)^2$

"N" indicated the number of p-channel FETs in saturation

$A_{ns}(N) := \text{find}(V_{in}, V_{ds1})$

$N := 1 .. 2$

$A_{ns}(N)_0$	$A_{ns}(N)_1$
2.345	0.175
2.622	0.245

The lowest single input transition midpoint occurs at an input voltage of 2.345 V while the simultaneous double transition occurs at an input voltage of 2.622 V. Slight alteration of the program to model a single transition of input B (Equations (16.54) and (16.55) need slight modification) results in an input transition voltage of 2.388 V.

Transmission Gates

A transmission gate has an output signal that duplicates its input signal when a third signal, the ENABLE signal, is present. When the ENABLE signal is in its other state (often called the INHIBIT state) the transmission gate is opened. A very simple CMOS realization of a transmission gate is shown in Figure 16.26. The transmission path is A to C, and the ENABLE/INHIBIT signal is applied with opposite polarity to the two MOSFET gate terminals. Digital transmission gates are very similar the parallel CMOS analog switch discussed in Section 4.5 (Book 1), but optimized for single-direction transmission.

Transmission gates are often used with a clock signal entering the ENABLEterminal. As such, the transmission gate serves to gate signal on or off. These gates are commonly found in multiplexers and other digital devices requiring signal switching.

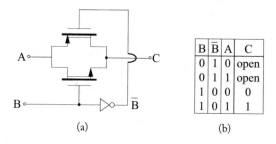

B	$\bar{B}$	A	C
0	1	0	open
0	1	1	open
1	0	0	0
1	0	1	1

(a) (b)

Figure 16.26: CMOS digital transmission gate: (a) Simplified circuit diagram; (b) Truth table.

16.5 BISTABLE LOGIC CIRCUITS

Electronic circuits constructed in such a manner so that two distinct, stable output states exist are commonly identified with the term *bistable*. Bistable logic circuits perform many important functions in digital circuitry. Level latches, counters, shift registers, and memories all depend on bistable circuitry. The most significant property of a bistable circuit is its ability to maintain a given stable output until an external input is applied. Application of an appropriate external input will cause a bistable circuit to change state in a predictable manner: the bistable circuit then holds the new output state until another appropriate input is applied. The most significant examples of bistable circuits are:

- Latches

- Flip-flops

- Schmitt Triggers

A *latch* is the simplest form of bistable circuit. This circuit "latches" its output to be the same logic level as its last valid input. The latch then holds the output at the logic level until another valid input forces a change in the latch state. As such, the latch is a very simple form of memory circuit. Latches are especially significant on shared data busses where values must be held while the buss transmits other data. *Flip-flops* are typically derived from latches. Most significant among the changes is the requirement that a clock pulse be present in order for a flip-flop to switch states. While many latches have an indeterminate output state, the output of a flip-flop is always determinate. The output of a flip-flop depends not only on the inputs but also on the current state of its output: in that sense, it also is a form of memory circuit. The *Schmitt trigger* finds greatest use in speeding the rise and fall times of digital signals that, for various reasons, have level transitions that are too slow for accurate logical manipulations. It is characterized by an input/output characteristic displaying hysteresis.

Detailed presentation of bistable circuitry is beyond the scope of this discussion. Only a few common circuits are presented for demonstration of the principles.

The SR Latch

The set-reset (SR) latch is a very common form of single-bit retention circuit. The SR latch is formed by cross-coupling the outputs of a pair of NOR gates into the inputs of the opposite member of the pair as shown in Figure 16.27. The output, Q, of the latch transitions to match the S input when only one input is HIGH. When both inputs are LOWthe latch retains its last value of Q and holds it until at least one input transitions to HIGH. The terms *set* and *reset* refer to the action of the output, Q. Q *sets* (transitions to HIGH) when the set input, S, is HIGH: Q *resets* (transitions to LOW) when the reset input, R, is HIGH. Unfortunately the simple SR latch has a state that must be avoided: when *both inputs* are HIGH the output is indeterminate.

A common realization of an SR latch using two NOR gates is shown in CMOS form is Figure 16.28.

Other IC gate families may use a different realization of the latch. In particular, it is more efficient in I^2L gate realizations to use a NAND form of the SR latch (Figure 16.29). This NAND realization produces the same logic characteristic as the NOR realization.

The JK Flip-Flop

Flip-flops are an augmentation of a basic latch[16] that removes the indeterminate state present when both inputs are HIGH. One common flip-flop is the JK flip-flop, logically realized in

---

[16]Terminology has not been effectively standardized for bistable circuitry. Many sources prefer to consider the SR latch a primitive form of flip-flop, however IC terminology usually reserves the term for the more complex circuitry.

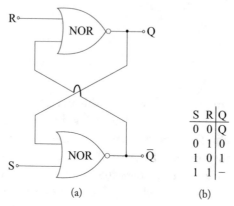

S	R	Q
0	0	Q
0	1	0
1	0	1
1	1	−

(a)   (b)

Figure 16.27: Basic SR latch: (a) Logic diagram; (b) Characteristic table.

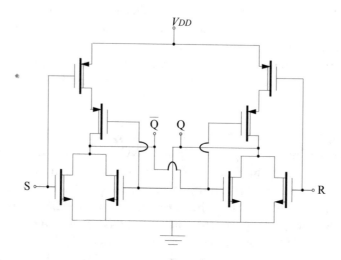

Figure 16.28: CMOS SR latch (NOR realization).

Figure 16.30a. The addition of two three-input AND gates and an addition feedback path removes the ambiguity in the logic table so that if both inputs are HIGH the output, Q, inverts. The addition of a clock signal avoids many of the problems associated with noisy input signals.

A JK flip-flop will only transition between states during the presence of a clock pulse. During that clock pulse the inputs and outputs are combined to form the logic table of Figure 16.30b. This table is the same as for a SR latch with the single exception that two HIGH inputs toggle the output to its complement in a JK flip-flop where that state resulted in an indeterminate state in the SR latch.

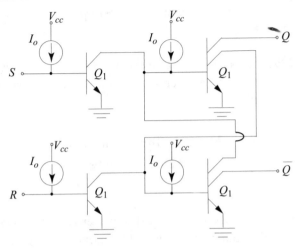

Figure 16.29: I^2L SR latch (NAND realization).

A characteristic of all JK flip-flops is that the output will toggle (change to the opposite state) when clocked in the presence of a HIGH signal at both inputs. Operated in that mode the circuit becomes a *T flip-flop* and is particularly useful in digital counters.

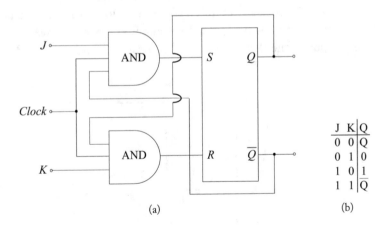

Figure 16.30: Basic JK flip-flop: (a) Logic diagram; (b) Characteristic table.

Since latches and flip-flops have standard logic gates as basic functional components, they are subject to many of the same speed restrictions: gates with short delay times lead to fast latches or flip-flops. Typically, the speed of a latch or flip-flop is specified through the *maximum clock frequency*. The maximum clock frequency is simply the highest rate at which the clock input of a bistable circuit can be driven while maintaining proper operation. Other significant operational parameters for these circuits are:

- *Setup and Hold times*: It is necessary that the input data arrive a short time before the triggering edge of the clock pulse and remain a short time after: these times are the setup time and hold time, respectively.

- *Clock* HIGH *and* LOW *pulse widths*: The minimum time that the clock must remain in its HIGH and LOW states for proper gate operation.

The Schmitt Trigger

The output of a Schmitt trigger is bistable and has very steep transition regions. The characterizing feature of the Schmitt trigger transfer function is the presence two separate transition regions, one for positive slope and one for negative slope signals, separated by a deadband region. The resultant transfer relationship exhibits hysteresis as shown in Figure 16.31. Any input signal below the negative slope transition voltage, V_T^-, results in a LOW output, V_L. If the output state is LOW, it will not transition to the HIGH state unless the input is greater than the positive slope transition voltage, V_T^+. Similarly, any input above V_T^+ results in a HIGH output, V_H, that will not transition to the LOW state unless the input falls below V_T^-. Thus, signals, after crossing a threshold, do not respond to input signal changes unless the variation is large enough to cross the deadband.

The Schmitt trigger is especially useful in converting slowly varying or a noisy signal into a clean digital form with sharp transitions. Another common usage is converting sine-wave input into a pulse-train output.[17] The dependence of the output on both level and level derivative is unique.

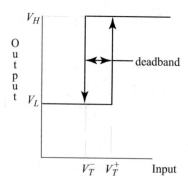

Figure 16.31: Schmitt trigger transfer characteristic hysteresis.

---

[17]Analog Schmitt trigger applications are presented in Section 13.1.

As with all bistable circuits, realizations of Schmitt triggers are possible using both FETs and BJTs. Two common realizations are shown in Figure 16.32. Similar circuits are available in most transistor gate families.

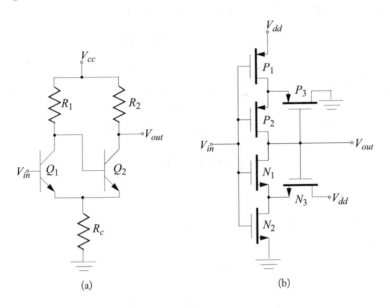

(a)   (b)

Figure 16.32: Typical Schmitt trigger circuits: (a) Bipolar; (b) CMOS.

Example 16.6

The Bipolar Schmitt Trigger Circuit of Figure 16.32a is constructed with BJTs described by $\beta_F = 50$ and component values:

$$V_{CC} = 5\,\text{V} \quad R_1 = 3.5\,\text{k}\Omega \quad R_2 = 2.6\,\text{k}\Omega \quad R_e = 1\,\text{k}\Omega.$$

Determine the Trigger voltages for positive and negative slope signals, V_T^+ and V_T^-.

Solution

When V_{in} is LOW, Q_1 is cut-off and Q_2 is in saturation. Two loop equations can be written:

$$5 - 3.5\,\text{k}\,I_{b2} - 0.8 - 1\,\text{k}\,(I_{b2} + I_{c2}) = 0,$$

and

$$5 - 2.6\,\text{k}\,I_{c2} - 0.2 - 1\,\text{k}\,(I_{b2} + I_{c2}) = 0.$$

The solution to this pair of equations is

$$I_{b2} = 679\,\mu\text{A} \quad \text{and} \quad I_{c2} = 1.145\,\text{mA}.$$

Which implies

$$V_{out} = 2.024\,\text{V} \quad \text{and} \quad V_{e2} = 1.824\,\text{V}.$$

The change of state will begin when Q_1 begins to turn on

$$V_T^+ \approx V_{e2} + V_{\gamma 1} = 1.824 + 0.5 = 2.33\,\text{V}.$$

When V_{in} is HIGH, Q_1 is in saturation and Q_2 is cut-off: $V_{out} = 5\,\text{V}$. The low threshold voltage can be found by determining when Q_2 begins to turn on. For this to happen, Q_1 must enter the forward-active region and have a collector-emitter voltage equal to the cut-in voltage of $Q_2 = V_{\gamma 2}$. The collector-emitter voltage of Q_1 (with Q_2 OFF) is given by:

$$5 - 3.5\,\text{k}\,\frac{\beta_F}{\beta_F + 1} I_{e1} - 1\,\text{k}\,I_{e1} = V_\gamma = 0.5.$$

The solution to this equation is

$$I_{e1} = 1.015\,\text{mA} \qquad V_T^- = V_{in} = 1\,\text{k}\,I_{e1} + 0.7 = 1.72\,\text{V}.$$

16.6 SEMICONDUCTOR MEMORIES

Memories are devices that are capable of the storage and retrieval of large amounts of digital data often required in modern digital systems. A wide variety of memory systems are available:

- Semiconductor memories

- Magnetic core memories

- Magnetic bubble memories

- Moving surface memories (magnetic tape, disc, etc.)

Discussion here will be focused on the first of these possibilities: semiconductor memories. In addition, the discussion will be restricted to *random-access memory* (RAM) and exclude *read-only memory* (ROM).

Semiconductor RAM is typically composed of large arrays of identical, single-bit memory cells that are accessed through row and column decoding electronics. The general topology for such an array is shown in Figure 16.33. Each cell must be capable of reading data from an external source, storing the data until needed, and then writing the data to an external element. Consequently, the memory array includes both address/command lines as well as data lines. Since memory addressing is based on binary codes, the number of rows and columns are integral powers of 2.

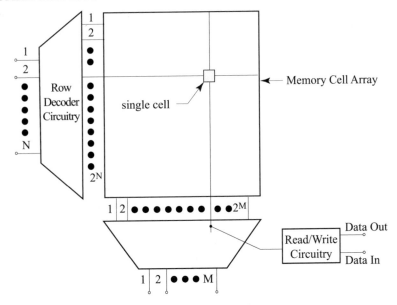

Figure 16.33: Semiconductor RAM array topology.

The topology of the individual memory cells distinguishes the various types of semiconductor RAM. A RAM cell typically stores information in either a digital latching circuit or in the charge on a capacitor. Latches can hold their state indefinitely (assuming no loss of electrical power), while the charge in a capacitor-based storage cell gradually dissipates and must be periodically refreshed by external circuitry. The term *static random access memory* (SRAM) applies to latch-based cells, while the term *dynamic random access memory* (DRAM) applies to cells that must be refreshed.

Static Random Access Memory
Typical static memories are based upon the simple digital latches discussed in Section 16.5. While any of the bipolar or MOS latch realizations discussed can be used to form static memories, the MOS realizations are currently the dominant technology. MOS SRAM memory cells are characterized by high speed, low power consumption, and high reliability. A six FET (6-T) CMOS memory cell based on a CMOS SR latch is shown in Figure 16.34. In this cell, input data, D_I, enters the latch through the FET switch formed by Q_5 and exits the latch through Q_6 as output data, D_O. A HIGH signal on the Read/Write line activates the two FET switches, enabling the entrance or exit of data. A LOW signal on the Read/Write line disconnects the latch from external circuitry: without any external stimulation, the latch holds its current state. As with all CMOS circuitry, the current flow in this cell is extremely low except when the cell is in transition

between memory states. The resultant low power consumption is extremely attractive for memory systems dependent on limited capacity power sources such as batteries.

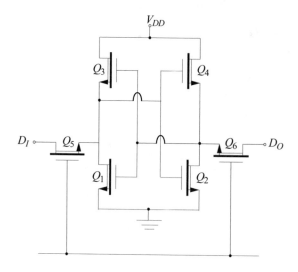

Figure 16.34: 6-T SRAM CMOS memory cell.

An NMOS version of the 6-T cell is also available. In the NMOS cell, Q_3 and Q_4 are replaced by active loads, typically enhancement-mode FETs. The NMOS realization of the 6-FET memory cell draws a more-consistent, albeit higher, current from the power supply. The price of less electrical noise than the CMOS realization is increased power consumption.

Another variation of this cell is the four FET (4-T) SRAM cell. This variation is essentially the same circuit as the NMOS realization. Here the active load FETs are replaced by polysilicon resistors. The variation allows for smaller cell size, but increases power consumption and decreases reliability somewhat over the other two realizations.

Dynamic Random Access Memory

In many situations, the packing density of memory cells can be an important factor in memory design. The dependence of static RAM on latches prohibits a size reduction beyond the 4-T size. In addition, the control lines to a single SRAM cell can be as many a six. Dynamic RAM can break these barriers by basing data memory on capacitor charge storage. The disadvantage of DRAM lies in the need to refresh the capacitor charge before leakage can degrade the data storage process. The need to keep the memory-storage capacitors small in size results in a refresh process cycling every 2 to 4 ms. Even though this seems frequent, for very large memory arrays, DRAM can have a distinct advantage over SRAM.

Two basic memory cells dominate DRAM design: a three FET (3-T) and a single FET (1-T) design. The 3-T design is shown in Figure 16.35: it is basically a capacitor, two switches, and an output buffer. Incoming data, D_I, enters through switch Q_1 and in stored in the capacitor. Q_2 acts as an inverting buffer so that the capacitor is not significantly discharged when the output switch, Q_3, is activated to write the output, D_O. The inversion of the output signal due to the buffer is usually compensated in either the read or write path with a single inverter. The 3-T cell requires separate read and write lines, but the data lines may be combined at the expense of more complex data encoding/decoding electronics.

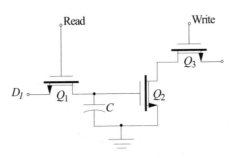

Figure 16.35: Typical 3-T DRAM memory cell.

The simplest of all DRAM cells is the 1-T cell shown in Figure 16.36. Here the data storage capacitor is connected to a single data line through a single switch. Incoming data charges the capacitor through the switch. When the data is to be read, the same switch connects the capacitor to the data line. This connection completely discharges the capacitor: it is then necessary to immediately refresh that data by imposing an amplified duplicate back onto the data line. Of course this immediate-refresh-after-read is in addition to the normal refresh process necessary in all DRAM. The cell size reduction possible in the 1-T warrants, in many situations, the added complexity of the drive electronics.

16.7   GALLIUM ARSENIDE LOGIC

Gallium Arsenide (GaAs) integrated circuits have a distinct speed advantage over similarly sized Silicon ICs. Gate operating speeds in excess of 1 Gigabit per second are currently available and the speed limit appears not to have been reached. In addition, GaAs circuits typically dissipate less power than Silicon circuits. Unfortunately, only two logic functions, an inverter or a NOR gate, can currently be constructed as a single logic stage: all more complex functions must be created from arrays of gates. In addition, the processing technology of GaAs circuits is not at the same stage of maturity as Silicon IC processing. Currently, GaAs wafers are significantly smaller than Silicon wafers. It is also currently difficult to produce transistors within a chip with uniform electrical properties compared to Silicon MOS device variation. This parameter variation, coupled

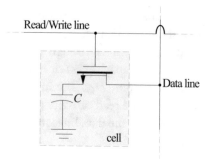

Figure 16.36: Simplified 1-T DRAM memory cell.

with a relatively high thermal coefficient for GaAs and relatively high power dissipation, places severe restrictions on noise margins and the reliability of logical operations. Still, it appears that GaAs circuits have a significant future.

Since a native oxide of GaAs does not exist, MOS-like structures are not possible in GaAs ICs. Therefore, other transistor structures have been developed. Three basic transistor structures have currently been shown to be useful in GaAs circuits:

- Metal-Semiconductor Field Effect Transistors (MESFETs),

- Heterojunction Field Effect Transistors (HFETs), and

- Heterojunction Bipolar Transistors (HBTs).

MESFETs are the current dominant GaAs transistor structure. Operation of a MESFET is similar to a JFET where a metal-semiconductor junction takes the place of the p-n junction of the JFET. As seen in Section 16.2, appropriate metal-semiconductor junctions form Schottky barrier diodes. Interestingly, MESFETs can be fabricated as either depletion-mode or enhancement-mode FETs. Voltage current relationships for GaAs MESFETs are essentially the same as for Silicon JFETs with parameters in the ranges:

$$-2.5\,\text{V} < V_P < -0.2\,\text{V} \quad \text{or} \quad 0.1\,\text{V} < V_T < 0.3\,\text{V}.$$

Typical GaAs MESFET Schottky barrier voltage is in the range of 0.8 V. HFETs are characterized by voltage-current relationships similar to MESFETs and are therefore not specifically discussed. HBTs are functionally similar to bipolar junction transistors. The discussion here is limited to MESFET gate structures.

There are two basic families of GaAs MESFET logic circuits:

- Enhancement-Depletion Logic, and

- Source-Coupled Logic.

The enhancement-depletion family bears considerable similarity to NMOS logic, and source-coupled logic is similar to the bipolar ECL. Because of these similarities, discussion here will be descriptive, rather than quantitative, in scope. Dominance among the GaAs logic families is not yet firmly established.

Enhancement-Depletion Logic

GaAs Enhancement-depletion (ED) logic circuits share the same topology with NMOS logic circuits. As an example, the topology of the GaAs NOR gate shown in Figure 16.37 is essentially the same as the NMOS NOR gate shown in Figure 16.22b. In this realization, Q_1 and Q_2 are enhancement-mode FETs and Q_3 is a depletion-mode FET used as an active load.[18] However, the functional differences between a GaAs MESFET and a Silicon MOSFET restricts direct comparisons. In particular, the Schottky barrier diode inherent to the gate of a GaAs device allows gate current to flow when gate-source voltages exceed the Schottky barrier voltage. Consequently, voltage swing must be kept small in order to avoid this detrimental gate current condition.

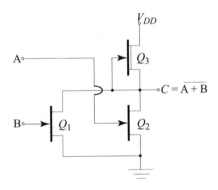

Figure 16.37: Simple GaAs enhancement-depletion NOR gate.

The principle differences between GaAs ED logic and NMOS logic are:

- For a HIGH output, any slave gate attached to the output will clamp the output voltage to the Schottky barrier voltage (≈ 0.8 V). The output voltage in NMOS is limited to the much larger value, $V_{DD} - V_T$.

- For a HIGH output, the current in the active load is not zero: power is dissipated for both HIGH and LOW output levels. NMOS dissipates essentially zero power for a HIGH output.

[18]The NMOS gate in Figure 16.22b uses an enhancement-mode FET as an active load rather than a depletion-mode FET. While GaAs requires a depletion-mode active load, NMOS can be fabricated in either form: operation of the gate is the same.

- When the input is HIGH, the output is LOW unless the input voltage exceeds the Schottky barrier voltage. Further increases in the input voltage increase the output voltage. If sufficiently high input voltage is applied, logical errors will occur. NMOS does not have this potential problem.

- Voltage swings are limited to the Schottky barrier voltage (≈ 0.8 V). NMOS swings can be much larger.

- The power supply voltage for GaAs ED logic needs to be only slightly larger than the Schottky barrier voltage. This supply voltage can therefore be significantly smaller than that of NMOS logic.

- The threshold voltage for GaAs enhancement FETs must be less than the turn-on voltage of the Schottky diode. NMOS FETs do not have this restriction.

While it is possible to construct a GaAs NAND gate using the same topology as a NMOS NAND gate, the very small differences in voltage levels between a logic HIGH and LOW reduce the noise margin to unacceptable levels. The "stacking" of enhancement FETs to create the logic NAND function increases the logic LOW and is responsible for this reduction in noise margins. GaAs NAND gates are not commercially available.

Source-Coupled Logic

GaAs source-coupled (SC) logic is based upon a FET differential amplifier[19] in the same manner as Emitter-coupled logic is based on a BJT differential amplifier. A simple two-input OR gate is shown in Figure 16.38. If either input, A or B, is a logic HIGH, Q_3 will enter cut-off and the output, C, will become a logic HIGH. Conversely, only if both inputs are a logic LOW will Q_3 enter saturation and the output goes to a logic LOW. The logic voltage levels for an unloaded gate are:

$$V_{oH} \approx V_{DD} - V_T \quad \text{and} \quad V_{oL} \approx V_T.$$

This logic swing is significantly larger than other forms of GaAs logic. Noise margins are also significantly improved.

One particularly attractive feature of SC logic is its insensitivity to transistor parameter variation. Unfortunately, SC logic consumes significantly more power than other forms of GaAs logic. A simple SC OR gate is roughly twice the size of the same gate in GaAs ED logic. Gate speed is essentially equivalent in the two logic families.

16.8 CONCLUDING REMARKS

The *essential aspects* of the electronic operation of the major digital circuits commonly in use have been presented in this chapter. Discussion has been limited to the basic building blocks of each

[19]Source-coupled differential amplifiers are discussed in Section 6.3 (Book 2).

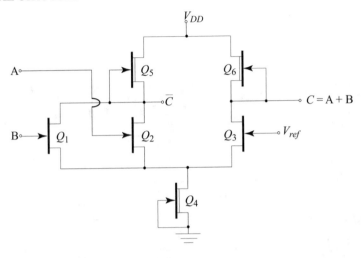

Figure 16.38: Simple GaAs source-coupled OR/NOR gate.

of the major logic families commonly in use. A short comparison of a four commercial logic gate families is shown in Table 16.1.

Summary Design Example

An existing TTL (LS) digital circuit is to be interfaced to a new system using CMOS (HC) digital gates. Both systems are to be operated from a 5 V power source. The nominal input and output specifications of the two logic families for 5 V operation are shown in Table 16.2.

The small difference in propagation delay presents no significant problems in design of an interface between the two circuits. There is, however a mismatch in acceptable logic levels. The CMOS output logic levels are compatible with the TTL input levels, but the TTL HIGH output ($V_{OH(min)} = 2.7\,\text{V}$) is not compatible with the CMOS HIGH input requirements ($V_{IH(min)} = 3.15\,\text{V}$). Interface circuitry to alleviate this incompatibility is to be designed.

Solution:

The output level of LS TTL gates is determined by the active pull-up circuitry: the voltage drop across this circuitry, even with minimal load current, does not allow the output voltage to *consistently* rise above the levels necessary for HCMOS circuitry. A variety of possible interfaces can be considered:

- Buffer the TTL output with a TTL gate that has passive pull-up.

- Buffer the TTL output with an HCT gate (these are HCMOS gates specifically modified to accept TTL levels at the input.

- Operate the CMOS portion of the circuit at a lower voltage level, i.e., 3 V.

Table 16.1: Comparison of commercial logic gate properties

Logic Family	TTL		ECL	CMOS
Series	H	LS	100K	HC
Basic Gate	NAND	NAND	OR/NOR	NOR OR NAND
Fan-out	10	20	25	>50
Power Dissipation	22.5 mW	2 mW	40 mW	1.75 nW @1MHz
Noise Margin-Low	0.4 V	0.3 V	0.14 V	1.25 V
Noise Margin-High	0.4 V	0.7 V	0.14 V	1.25 V
Propagation Delay	6 ns (C_L = 25 pF)	9.5 ns (C_L = 15 pF)	0.75 ns (C_L = 50 pF)	8 ns (C_L = 50 pF)
Maximum Clock Frequency (D-flip-flop)	50 MHz	33 MHz	350 MHz	50 MHz

Table 16.2: Nominal input and output specifications of the two logic families for 5 V operations

	LS TTL	HC CMOS
$V_{OH(min)}$	2.7 V	4.4 V
$V_{IH(min)}$	2.0 V	3.15 V
$V_{OL(max)}$	0.5 V	0.1 V
$V_{IL(max)}$	0.8 V	0.9 V
$I_{OL(max)}$	8 mA	4 mA
$I_{IL(max)}$	−0.4 mA	−1 μA
$I_{OH(max)}$	−0.4 mA	−4 μA
$I_{IH(max)}$	20 μA	1 μA
Propagation Delay	9 ns	8 ns

• Shunt the TTL active pull-up with an external, passive pull-up.

Buffering can provide a good solution. The only significant drawbacks are an addition of components and the additional propagation delay induced by the buffer. TTL gates with passive pull-up may introduce a large additional propagation delay and are, therefore discarded as an alternative. If the new CMOS system can be implemented with HCT gates as the input gates, this is probably the best solution. Operation of the CMOS portion of the circuit at 3 V will lower the required HIGH input voltage requirement below the TTL output HIGH. Unfortunately this also reduces noise margins and violates the constraints in the design requirements. If HCT input gates are not possible, shunting the TTL active pullup with an external, passive pull-up seems a viable design alternative.

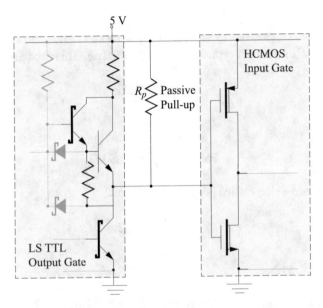

An external resistor that shunts the output of the LS TTL gate will force the HIGH level at the interface near 5 V. The minimum value of the pull-up resistor, R_p, is determined by the current sinking capability of the TTL gate: the LOW voltage must remain within specifications. For a LOW, the maximum current through the resistor is given by the sum of the maximum TTL output current and N CMOS input currents. This LOW voltage requirement restricts R_p to:

$$R_p > \frac{V_{CC} - V_{OL(max)}^{TTL}}{I_{OL(max)}^{TTL} - N I_{IL}^{HC}} \approx 563 \ \Omega.$$

The maximum value of R_p is determined by the output voltage rise time. This voltage rise is a complex process. It rises to $V_{OH(min)}$ very quickly (≈ 9 ns) due to the active pull-up. Thereafter, the rise will be exponential due to the RC time constant formed by the input capacitance of the

CMOS gates and R_p. The time period for the voltage to exceed the CMOS $V_{IH(min)}$ is given by the solution to:

$$v(t) = V_{CC} - \left(V_{CC} - V_{OH(min)}^{TTL}\right) e^{-\frac{t}{R_p C}} = V_{IH(min)}^{HC}$$

$$v(t) = 5 - 2.3\, e^{-\frac{t}{R_p C}} = 3.15 \quad \Rightarrow \quad t = 0.22 R_p C.$$

If a design using a shunt, passive pull-up resistor is to be better than a design using a buffer gate, the additional rise time must be shorter than the propagation delay due to a buffer. That is:

$$t < 9\,\text{ns} \quad \Rightarrow \quad R_p < \frac{(9\,\text{ns})}{(0.22)\,C} = \frac{40.9\,\text{ns}}{C}.$$

The total capacitance, C, is given by the input capacitance of the total number of CMOS gates being driven by the TTL gate. A typical value is: $C \approx 10\,\text{pF}$. This capacitive assumption leads to:

$$R_p < 4.09\,\text{k}\Omega.$$

Thus,

$$563\,\Omega < R_p < 4.09\,\text{k}\Omega.$$

A reasonable, standard-value choice is

$$R_p = 1.5\,\text{k}\Omega.$$

16.9 PROBLEMS

16.1. The BJT in the simple inverter shown is described by:

$$\beta_F = 75 \quad f_T = 200\,\text{MHz} \quad \beta_R = 2.$$

If the input voltage, V_{in}, has logic levels, 0 V and 5 V, determine the following:

(a) The rise time

(b) The delay time

(c) The fall time

(d) The storage time

(e) The average propagation delay

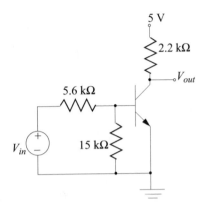

16.2. Assume the power supply for the simple inverter of Problem 16-1 is reduced to 3.3 V and the input logic levels become 0 V and 3.3 V. The transistor parameters are unchanged. Determine the following:

(a) The rise time

(b) The delay time

(c) The fall time

(d) The storage time

(e) The average propagation delay

16.3. In an attempt to decrease the propagation delay of the simple inverter described in Problem 16-1, the collector resistor is reduced to 1.5 kΩ. Comment *quantitatively* on the advisability of this design change.

16.4. The BJTs in the inverter with active pull-up shown are described by:

$$\beta_F = 75 \quad f_T = 200 \,\text{MHz} \quad \beta_R = 2.$$

If the input voltage, V_{in}, has logic levels, 0 V and 5 V, determine the following:

(a) The rise time

(b) The delay time

(c) The fall time

(d) The storage time

(e) The average propagation delay

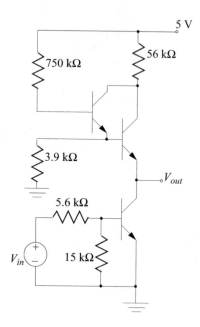

16.5. For the simple bipolar inverter shown:

(a) Use SPICE to determine the gate speed. Assume the BJT has the same properties as a 2N2222 NPN bipolar transistor.

(b) Model a Schottky inverter by shunting the base-collector junction of the given BJT with a diode. Repeat the SPICE gate speed determination of part a). Assume the diode has properties

$$I_s = 2\,\text{nA} \qquad \eta = 2.1.$$

(c) Comment on any gate speed differences.

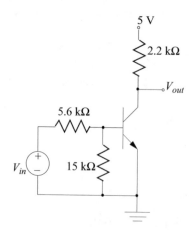

16.6.   The BJTs in the inverter with active pull-up shown are described by:

$$\beta_F = 75 \quad f_T = 200\,\text{MHz} \quad \beta_R = 2.$$

If the input voltage, V_{in}, has logic levels, 0 V and 5 V, determine the following:

(a) The rise time

(b) The delay time

(c) The fall time

(d) The storage time

(e) The average propagation delay

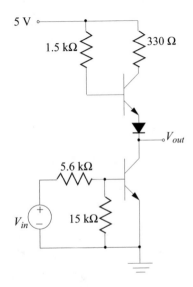

16.7.   The BJTs in the inverter with active pull-up and pull-down as shown are described by:

$$\beta_F = 75 \quad f_T = 200\,\text{MHz} \quad \beta_R = 2.$$

If the input voltage, V_{in}, has logic levels, 0 V and 5 V, determine the following:

(a) The rise time

(b) The delay time

(c) The fall time

(d) The storage time

(e) The average propagation delay

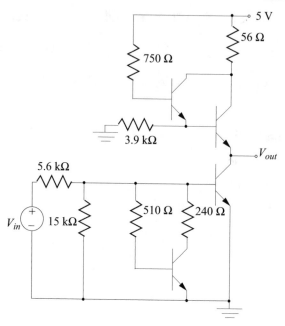

16.8. The circuit shown is a form of simplified TTL gate.

(a) Determine the logical operation the gate performs on the four inputs: A, B, C and D.

(b) What are the logic voltage levels at the output, V_{out}?

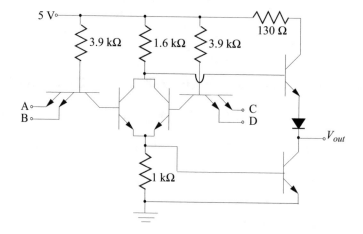

16.9. Compare the average power consumption of the simple ECL OR gate of Figure 3.12 (Book 1) to that of a 10 K OR gate (Figure 16.13) with the following parameters:

$$R_{C2} = 218\,\Omega \quad R_{C3} = 246\,\Omega \quad R_{EP} = 777\,\Omega$$
$$R_1 = 909\,\Omega \quad R_2 = 6.12\,k\Omega \quad R_3 = 4.99\,k\Omega \quad R_{IN1} = R_{IN2} = 51\,k\Omega.$$

Assume a pull-down resistor at the output of 1.5 kΩ and BJTs with $\beta_F = 100$.

16.10. A CMOS inverter is fabricated using a 3.3 V supply and MOSFETs with the following properties:

$$K_N = 0.1\,\text{mA/V}^2 \quad K_P = 0.25\,\text{mA/V}^2 \quad \text{and} \quad |V_T| = 1.$$

(a) Determine the average propagation delay time if it is driving a capacitive load of 5 pF.

(b) Compare the results of part a to those found in Example 16.2.

16.11. A CMOS inverter is fabricated using a 5 V supply and MOSFETs with the following properties:

$$K_N = 0.4\,\text{mA/V}^2 \quad K_P = 0.8\,\text{mA/V}^2 \quad \text{and} \quad |V_T| = 1.2.$$

Determine the average propagation delay time if the gate is driving a capacitive load of 5 pF.

16.12. A CMOS inverter is fabricated using a 3 V supply and MOSFETs with the following properties:

$$K_N = 0.06\,\text{mA/V}^2 \quad K_P = 0.15\,\text{mA/V}^2 \quad \text{and} \quad |V_T| = 0.6\,\text{V}.$$

Determine the average propagation delay time if the gate is driving a capacitive load of 15 pF.

16.13. A CMOS inverter is fabricated using transistors with the following properties:

$$K_N = K_P = 0.1\,\text{mA/V}^2 \qquad |V_T| = 1.5\,\text{V}.$$

Use SPICE to determine the voltage transfer characteristic for the following power supply conditions:

(a) $V_{DD} < V_T$

(b) $V_{DD} = V_T$

(c) $V_{DD} = 2V_T$

(d) $V_{DD} = 3V_T$

Comment on the results.

16.14. Given a 5 V power supply. Design a three-input NMOS NOR gate to have the following output logic levels when driven by a gate of the same design:

$$V_{oH} = 4.0\,V \qquad V_{oL(max)} = 0.8\,V.$$

What are the values of V_{oL} for one, two and three inputs HIGH?

16.15. Given a 3.3 V power supply. Design a three-input NMOS NOR gate to have the following output logic levels when driven by a gate of the same design:

$$V_{oH} = 2.7\,V \qquad V_{oL(max)} = 0.7\,V.$$

What are the values of V_{oL} for one, two and three inputs HIGH?

16.16. Given a 5 V power supply. Design a three-input NMOS NAND gate to have the following output logic levels when driven by a gate of the same design:

$$V_{oH} = 4.0\,V \qquad V_{oL(max)} = 0.8\,V.$$

16.17. Given a 3.3 V power supply. Design a three-input NMOS NAND gate to have the following output logic levels when driven by a gate of the same design:

$$V_{oH} = 2.7\,V \qquad V_{oL(max)} = 0.7\,V.$$

16.18. Determine a logical expression for the output, Y, of the CMOS circuit shown as a function of the three inputs, A, B, and C.

Use SPICE to verify the logical expression. Assume the MOSFETs are described by:

$$|V_T| = 1\,V \qquad K = 0.4\,mA/V^2.$$

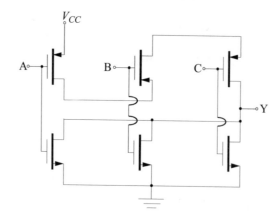

16.19. Given a 3 V power supply. Design a three-input NMOS NAND gate to have the following output logic levels when driven by a gate of the same design:

$$V_{oH} = 2.0\,V \qquad V_{oL(\text{max})} = 0.5\,V.$$

16.20. Determine the range of input voltages at which the midpoint of a logic transition occurs for a two-input CMOS NAND gate that uses FETs described by:

$$|V_T| = 0.5\,V \quad \text{and} \quad K_n/K_p = 2,$$

and a 3 V power supply.

16.21. Determine the range of input voltages at which the midpoint of a logic transition occurs for a two-input CMOS NAND gate that uses FETs described by:

$$|V_T| = 1.0\,V \quad \text{and} \quad K_n/K_p = 4,$$

and a 5 V power supply. Compare this voltage range to that found in Example 16.5.

16.22. It is possible to form bipolar gates that operate similarly to MOS gates. One such circuit is shown. Determine the logic function implemented by this circuit.

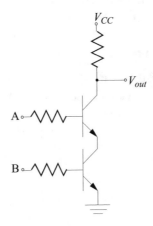

16.23. Use SPICE to implement the CMOS SR Latch of Figure 16.28. Assume a 5 V power supply and MOSFETs described by:

$$K_N = 0.1\,\text{mA/V}^2 \quad K_P = 0.25\,\text{mA/V}^2 \quad \text{and} \quad |V_T| = 0.6\,V.$$

Verify correct operation of the latch.

16.24. Design a bipolar Schmitt trigger using BJTs characterized by $\beta_F = 75$ and a 5 V power supply to meet the following design criteria.

Threshold voltages:

$$V_T^+ = 2.2\,\text{V} \quad \text{and} \quad V_T^- = 1.5\,\text{V}.$$

Power consumption:

Power supplied by the 5 V source must not exceed 5 mW in any stable state. Any current entering the input terminal of the Schmitt trigger is excluded from this calculation.

16.25. Design a bipolar Schmitt trigger using BJTs characterized by $\beta_F = 100$ and a 3.3 V power supply to meet the following design criteria.

Threshold voltages:

$$V_T^+ = 2.0\,\text{V} \quad \text{and} \quad V_T^- = 1.2\,\text{V}.$$

Power consumption:

Power supplied by the 3.3 V source must not exceed 3 mW in any stable state. Any current entering the input terminal of the Schmitt trigger is excluded from this calculation.

16.26. Use SPICE to determine the transfer characteristic for the given Silicon bipolar Schmitt trigger. The BJTs are characterized by $\beta_F = 50$. Compare SPICE results to those obtained by the simple hand analysis of Example 16.6. Over what range of input voltages are the two transition regions?

Hint: In transient analysis SPICE may have difficulty converging for switching circuits: this can usually be eliminated by setting ITL4 = 40 in a .OPTIONS statement.

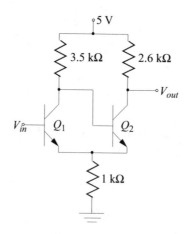

16.27. In the Schmitt trigger circuit shown, the BJTs are characterized by $\beta_F = 50$.

(a) Using simple hand analysis, determine the positive and negative slope trigger voltages, V_T^+ and V_T^-.

(b) Use SPICE to validate results obtained by the simple hand analysis.

Hint: In transient analysis SPICE may have difficulty converging for switching circuits: this can usually be eliminated by setting $ITL4 = 40$ in a .OPTIONS statement.

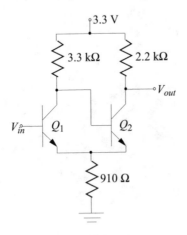

16.28. Use SPICE to determine the voltage transfer characteristic for the given MOSFET Schmitt trigger. The MOSFETs are characterized by:

$$V_T = 1\,\text{V} \qquad K = 0.2\,\text{mA/V}^2.$$

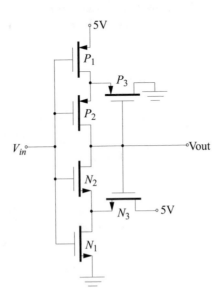

16.29. An existing TTL (LS) digital system is to be interfaced to a new ECL digital system. The interface is unidirectional with the TTL system driving the ECL system. It has been suggested that the interface can be realized using a common-base amplifier. The nominal input and output specifications of the two logic families are known to be:

	LS TTL	10K ECL
$V_{OH(min)}$	2.7 V	−0.98
$V_{IH(min)}$	2.0 V	−1.105
$V_{OL(max)}$	0.5 V	−1.85
$V_{IL(max)}$	0.8 V	−1.475
$I_{OL(max)}$	8 mA	22.3 mA
$I_{IL(max)}$	−0.4 mA	65 μA
$I_{OH(max)}$	−0.4 mA	5.4 mA
$I_{IH(max)}$	20 μA	130.5 μA
Propagation Delay	9 ns	2.5 ns

Complete the interface design and verify proper operation using SPICE.

16.30. It has been suggested that the TTL-CMOS interface described in the Summary Design Example could be improve upon using an active logic-level interface circuit. One such interface circuit is shown.

Choose appropriate resistor values and compare the operation of this interface circuit to that of the Summary Design Example. Assume BJTs described by:

$$\beta_F = 50.$$

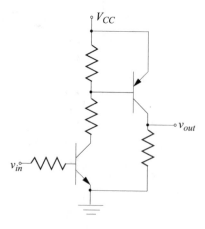

REFERENCES

[1] ——-, *High-Speed CMOS Logic Data Book*, Texas Instruments Inc., Dallas, 1988.

[2] Buchanan, James, *CMOS/TTL Digital Systems Design*, McGraw-Hill Book Company, New York, 1990.

[3] Glasford, Glenn, *Digital Electronic Circuits*, Prentice Hall, Inc., Englewood Cliffs, 1988.

[4] Haznedar, Haldun, *Digital Microelectronics*, The Benjamin/Cummings Publishing Company, Inc., 1991.

[5] Hodges, David and Jackson, Horace, *Analysis and Design of Digital Integrated Circuits, 2nd Ed.*, McGraw-Hill Book Company, New York, 1988.

[6] Shoji, Masakazu, *CMOS Digital Circuit Technology*, Prentice Hall, Inc., Englewood Cliffs, 1988.

[7] Taub, Herbert and Schilling, Donald, *Digital Integrated Electronics*, McGraw-Hill Book Company, New York, 1977.

[8] Wing, Omar, *Gallium Arsenide Digital Circuits*, Kluwer Academic Publishers, Boston, 1990.

Authors' Biographies

Thomas F. Schubert, Jr., and Ernest M. Kim are colleagues in the Electrical Engineering Department of the Shiley-Marcos School of Engineering at the University of San Diego.

THOMAS F. SCHUBERT, JR.

Thomas Schubert received BS, MS, and PhD degrees in Electrical Engineering from the University of California at Irvine (UCI). He was a member of the first engineering graduating class and the first triple-degree recipient in engineering at UCI. His doctoral work discussed the propagation of polarized light in anisotropic media.

Dr. Schubert arrived at the University of San Diego in August, 1987 as one of the two founding faculty of its new Engineering Program. From 1997–2003, he led the Department as Chairman, a position that became Director of Engineering Programs during his leadership tenure. Prior to coming to USD, he was at the Space and Communications Division of Hughes Aircraft Company, the University of Portland, and Portland State University. He is a Registered Professional Engineer in the State of Oregon.

In 2012, Dr. Schubert was awarded the Robert G. Quinn Award by the American Society of Engineering Education "in recognition of outstanding contributions in providing and promoting excellence in engineering experimentation and laboratory instruction."

ERNEST M. KIM

Ernest Kim received his B.S.E.E. from the University of Hawaii at Manoa in Honolulu, Hawaii in 1977, an M.S.E.E. in 1980 and Ph.D. in Electrical Engineering in 1987 from New Mexico State University in Las Cruces, New Mexico. His dissertation was on precision near-field exit radiation measurements from optical fibers.

Dr. Kim worked as an Electrical Engineer for the University of Hawaii at the Naval Ocean Systems Center, Hawaii Labs at Kaneohe Marine Corps Air Station after graduating with his B.S.E.E. Upon completing his M.S.E.E., he was an electrical engineer with the National Bureau of Standards in Boulder, Colorado designing hardware for precision fiber optic measurements. He then entered the commercial sector as a staff engineer with Burroughs Corporation in San Diego, California developing fiber optic LAN systems. He left Burroughs for Tacan/IPITEK Corporation as Manager of Electro-Optic Systems developing fiber optic CATV hardware and systems. In 1990 he joined the faculty of the University of San Diego. He remains an active consultant in radio frequency and analog circuit design, and teaches review courses for the engineering Fundamentals Examination.

Dr. Kim is a member of the IEEE, ASEE, and CSPE. He is a Licensed Professional Electrical Engineer in California.

Printed in the United States
by Baker & Taylor Publisher Services